Formal Development of a Network-Centric RTOS

Eric Verhulst • Raymond T. Boute
José Miguel Sampaio Faria • Bernhard H.C. Sputh
Vitaliy Mezhuyev

Formal Development of a Network-Centric RTOS

Software Engineering for Reliable Embedded Systems

 Springer

Eric Verhulst
Altreonic NV
Gemeentestraat 61AB1
B3210 Leuven, Belgium
Eric.Verhulst@lancelot.be

José Miguel Sampaio Faria
Rua Sra das Boas Novas 776
4935-490 Mazarefes
Portugal
jmfaria@criticalsoftware.com

Vitaliy Mezhuyev
Open License Society
Zavelstraat 160
3010 Leuven
Belgium
Vitaliy.Mezhuyev@openlicensesociety.org

Raymond T. Boute
Department of Information Technology
Universiteit Gent
Faculty of Engineering
St. Pietersnieuwstraat 41
9000 Gent
Belgium
boute@intec.UGent.be

Bernhard H.C. Sputh
Open License Society
Zavelstraat 160
3010 Leuven
Belgium
bernhard.sputh@openlicensesociety.org

ISBN 978-1-4899-9288-8   ISBN 978-1-4419-9736-4 (eBook)
DOI 10.1007/978-1-4419-9736-4
Springer New York Dordrecht Heidelberg London

Printed on acid-free paper

Springer is part of Springer Science+Business Media (www.springer.com)

Preface

How can one improve with a factor of 10 on something that has already the reputation of being highly optimised? The answer lies in ignoring the most often wrong assumption that it is already highly optimised and by going back to basics. This inevitably includes developing a new formalisation of the problem at hand. In our case, this meant thinking anew about what a distributed RTOS (Real Time Operating System) is all about. What is the core functionality of an RTOS, of a distributed RTOS? Is there a clean way to handle task synchronisation and communication? The result was the unique network-centric OpenComRTOS project described in this book.

Taking this as an opportunity, we wanted to use formal methods to prove the final implementation. It turned out that formal methods can help to prove an implementation, but they really shine when used to model the architecture at an abstract level before any implementation is done. Their use has shown us again how much we are all influenced by what we know. After all our brains have a hard time reasoning without prior knowledge. Hence, our brains tend to look for known patterns so that known rules can be applied.

Looking for better and new solutions is hampered by prior knowledge. Formal methods help us because they allow us (or some would say: force us) to think at a more abstract level, our vision being less cluttered by implementation details. The result obtained in the project was a very clean and scalable architecture while verification had almost become trivial. There is also a general assumption that trustworthy means complex and large. Great was the surprise, however, when we discovered it resulted in the opposite. The RTOS was measured to be up to 10 times smaller than a previously hand coded version that had been tweaked over several years and used in demanding systems. This means less resources and less power are needed. So, to make the world less energy-hungry, use formal methods.

This project has to some extent reinvented the very concept of what an RTOS is. It is a way to model, it is a way to simulate, it is a way to verify, it is a way to program in a scalable and portable way concurrent systems. But our quest does not stop.

OpenComRTOS is also an enabler for new functionality that is still being researched while the book is being written. A lot of the work has to do with researching the correct semantics to support e.g. composability, dynamic resource scheduling and fault tolerance. Ultimately, it might result in new hardware.

Last but not least, formal methods have proven not to be so hard to use as it was assumed to be. The project also demonstrated the strength of team work. Communication in a well working team is ultimately the way to get rid of the assumptions our brains involuntary make. Formal methods again help by replacing intuition by abstraction. This book is not an academic one. It describes aspects that were explored during a real industrial project to develop a distributed RTOS from scratch using formal methods. Therefore it contains as well a broad discussion on the context in which such RTOS are used, as well as deep technical details of some of the formal models used. But as such, the description is not complete because it describes a project, not a theory.

The book is organised as follows: In the first two chapters, we sketch the domain of interest: trustworthy embedded real-time distributed systems. We discuss the challenges to develop applications and systems in this domain and why formal methods are becoming essential tools for the engineer working in this field. We derive from it the requirements and specifications for OpenComRTOS. In the following two chapters we look at what formal methods and tools are available and introduce TLA+/TLC that was finally selected and used in the project. Subsequently, we discuss the formal TLA+ models, as well as the architecture, of OpenComRTOS. We dwell a bit deeper on the interaction semantics and provide an overview of the code size and performance results. For the interested user the appendix includes a usage tutorial, as well as the mathematical and logic foundations behind temporal logics like TLA+. The appendix also contains the TLA+ and SPIN models used to compare both formalisms in Chap. 3.

For the interested reader, a free version of OpenComRTOS for PC is available from www.altreonic.com. This version also acts as a simulator and cross development environment for multi-node targets.

Acknowledgements

This work has been made possible by the support of many people and organisations:

- Alexander Keda for developing the verification models and code generators.
- Anatoliy Konovalenko for developing the RTOS unit tests.
- Andrey Nitsenko for developing the graphical event tracer.
- Annie Dejonghe for moral support and administrative support.
- Bernhard Sputh for managing the release of the product and porting the RTOS.
- Dimitry Panfilov for developing the first visual front-end and porting the RTOS.
- Gjalt De Jongh for his conceptual discussions and first implementations.
- José Miguel Faria for developing the first formal models.

- Raymond Boute for his deep knowledge of formal techniques.
- Vitaliy Mezhuyev for his meta-modelling input.

The project was also financially supported by IWT of the Flemish Region and Melexis NV. Melexis also provided the first target processor.

Linden   Eric Verhulst

Contents

List of Figures

List of Tables

List of Tables

Part I
Trustworthy Embedded Systems

Chapter 1
Introduction: OpenComRTOS Role in a Unified Systems Engineering Methodology

OpenComRTOS is part of a systematic, formalised systems and software engineering methodology for embedded systems with a supporting environment and tools. While OpenComRTOS can be used independently of it, users will benefit from using the methodology in an integrated way. This methodology is characterised by two key concepts: unified semantics and interacting entities. When used in combination, they result in a better control of the engineering process leading to the development of systems and products. OpenComRTOS plays an important role in this approach as it is the system software layer allowing the mapping of the abstract interacting entities at the modeling level into concrete concurrent instances.

1.1 Introduction

Our economy, our social and political environment can be considered as a system of systems. As citizens, we want these systems to work for us and to improve our lives. Technology and engineering are playing a growing and important role in it. The main reason for this fact is that technology allows us to do more with less. Technology provides us with efficiency. The task of the engineer is to put technology at work and to develop systems and products that provide us with added value. This applies to many domains, even in domains where technology only plays a supporting role and the role of the human is still dominant.

The authors of this book are mostly concerned with the domain of so-called embedded systems. While there is no unique definition for this domain, think about it as the domain of devices and systems that have a processor and software inside, often fully invisible to the user. It came into being when the transistor was invented. This was the start of the digital electronics era. Digital implies that it became more and more practical for engineers to start building systems based on the concept of state machines. What the solid-state transistor changed was that because of its shrinking size, many of these components could be used together to build very

E. Verhulst et al., *Formal Development of a Network-Centric RTOS: Software Engineering for Reliable Embedded Systems*, DOI 10.1007/978-1-4419-9736-4_1, © Springer Science+Business Media, LLC 2011

large scale state machines. A typical example is the processor in a desktop PC, now containing several of such devices, each having close to a billion transistors. Even a small processor can contain a few 10,000 to a few 100,000 transistors. On top of that, engineers made these components programmable. This comes down to using components whose functionality changes all the time (essentially at the rate of their clocks, often measured in MHz or GHz). The programs they run are composed of elementary instructions, meaning that the use of programs increases the size of the state machine exponentially. How do we ensure that such systems can be trusted to be correct?

This is not an easy task. Before electronics, most systems were analogue or mechanical ones. Such systems often require a lot of energy and are bulky, but usually they are quite trustworthy. The reason for this is that such systems inherently provide what is called "graceful degradation". Their state space is continuous and hence infinite, but when the material properties are affected by e.g. wear and tear, a mechanical system will keep delivering its function, even when it will have become less efficient. This is the property of graceful degradation. Of course, at some state, the system will break down as well, but there will be ample warning (if one cares to look and listen).

Digital electronic systems are often designed and manufactured in such a way that each individual transistor remains in a safe domain over its anticipated lifetime, just like with mechanical system. The difficulty comes from the fact that in an electronic system, these transistors are connected and therefore they create a large state machine. When a single transistor or its connections to another transistor fails for some reason, the system might continue to work but there is also a non-zero probability that the failure will bring the whole system to a halt. Often this means it goes into an illegal, read: undefined, state. Fortunately, in (small) digital electronics the state space is still combinatorial and in principle, one can simulate the system across all these states or one can even design a test set-up that will exercise all possible states, allowing to verify that the design prevents the system from reaching such an illegal state, even if such an event is very unlikely under normal operating conditions. The issue is that reaching such an illegal state can become very likely when the operating conditions are no longer "normal" (e.g. because the external conditions put the device outside its normal operating conditions). Often, the result will be catastrophic.

The problem really becomes horrendous when we look at embedded software running on such an electronic component. The issue is that now the size of the state space is exponentially expanded. This is partly due to the way software instructions are encoded in the hardware. If a single bit is changed, the behaviour can become entirely different. In addition, programmable electronic components are often built as so-called von Neumann machines. The processor instructions are executed in sequence. The program will also contain branching points, meaning that the resulting state space can grow very large, even under normal operating conditions. Moreover, embedded software will often not have the property of graceful degradation. If for some reason the next instruction is not the right one, the system can come to a halt in nanoseconds and standard processors cannot recover from such errors. A hard reset and rebooting from the beginning is often the only sensible option. Most of us are familiar with this notion, often called a "blue screen", but very few know that an ordinary PC will have at least one memory bit flipped per day due to cosmic radiation. While this is often innocent, when such an event occurs in a safety critical system, lives can be at stake.

Given that the state space is now exponential and that it is physically impossible to test all possible states, how can we then have confidence in embedded software? The solution engineers adopt is to prove that the software will be correct (this holds under the assumption that the hardware is correct as well). This is essentially not different from what engineers do in other domains. For example, construction and material engineers will often not test their construction to see when it will fail. No, they will develop a mathematical model and calculate the breaking point based on the assumption that their raw materials were correctly manufactured. This allows them to apply a hefty safety margin to their design. Unfortunately, software cannot be made robust by adding somewhere a safety margin, hence we must "calculate" it exactly. This is what the emerging field of formal techniques is all about and this book is about its application to the development of a crucial embedded software component: a network centric Real-Time Operating System.

Another aspect is that the development of embedded software is not a "standalone" activity. Embedded application software has many dependencies, often on third party input or components. In addition, embedded software is essentially implementing a real-world context as a computer program. If the description of this real-world context is erroneous, these errors will be found back in the resulting application software and there they can result in erroneous products even if the implementation of the software was done correctly.

Therefore, we need to look at the whole systems engineering process. This is essential to develop trustworthy products because engineering a product involves a lot of human activity. It is a complex process with many aspects and many problems that need to be mastered. One of them is the use of natural language. Because natural language is not precise enough, often vastly differing between cultures and different domains, it is the source of many issues in systems engineering. Therefore, we must try to achieve a common language across all domains that are involved in the engineering of a product or a system. We called this trying to achieve "unified semantics". The only way to do this is to develop a unified "systems grammar" as we call it, that covers the full domain of systems (or software) engineering. This is similar to the development of an ontology but it adds the notion of "interaction" to make the relationships between the concepts concrete from early requirements to the final release of the product or system being developed.

Just like in a language it defines terms of a vocabulary and relationships between these terms. Such a systems grammar will also seek orthogonality, essentially trying to come up with terms that have no overlapping and no ambiguous meaning. Less is often better in this context. It can be understood as an application of Einstein's principle (or occam's razor if you prefer). Keep things simple, but not too simple. Essentially, if a solution is complex, it is not because its creators were smart, but because they did not fully understand the problem at hand. Below follows a description of the systems engineering methodology we developed and of which the OpenComRTOS project was at the same time a test case as well as an important key component of the supporting environment.

1.2   A Systematic Engineering Methodology Based on Unified Semantics and Interacting Entities

Generally speaking, systems engineering starts with what is called "requirements and specifications capturing". This is a domain where many people, called stakeholders, are involved. In a first instance they formulate, often more informally than formally, "requirements". Requirements are often not precise enough to serve as a basis for implementation. They will often be contradictory, overlapping and stretch across multiple, very divergent domains. To find out which requirements are really relevant and what they really aim for requires a lot of social interaction and discussions (Fig. 1.1).

Often, the term "requirement" is wrongly used to indicate what good systems engineering calls "specifications". A specification is defined as a quantified or qualified requirement statement. This means it is linked with measurable values, generally called verifiable properties. When we say "verifiable" we can design the system to achieve the specified properties. A consequence is that we also have to specify how we will measure, verify and test the specified properties as the methods we use will often influence the result. We call this the "test case". This is a first domain were standardization might be useful to reduce the guess work and to increase the repeatability of the process. Another aspect is that we need to specify the circumstances under which the specifications will be met. The fact that a specification is linked with measurable values indicates that there must be one or multiple test cases, which means that the design must take the test conditions into account. The test case will often only be present when testing but it will often require the design to have provisions for it.

In normal operating conditions, we speak of the "normal" case. But as we have seen earlier, undesired behaviour often happens in conditions that are unlikely and often outside the normal operating parameters. This is called the "fault case". Thinking about fault cases is difficult because it requires to think in reverse: "if the system fails in a certain way, what type of failure could that be and what chain of events could cause it?".

Next comes the step where the specifications are realized by a concrete design. Although often it will be required that the system is implemented using previously developed or reused components, the difficulty here is not to let the implementation choice influence the specifications. In the ideal world, the engineer will now build several "models". Models are essentially partial implementations that fulfil selected specified properties of the system. We distinguish here three classes. Simulation models are essentially virtual prototypes of parts or of the whole of the system as

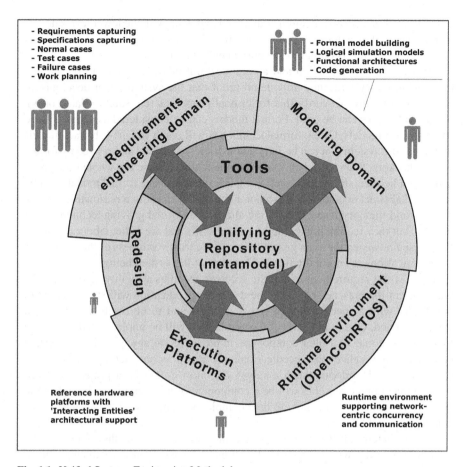

- Requirements capturing
- Specifications capturing
- Normal cases
- Test cases
- Failure cases
- Work planning

- Formal model building
- Logical simulation models
- Functional architectures
- Code generation

Requirements engineering domain

Modelling Domain

Tools

Redesign

Unifying Repository (metamodel)

Execution Platforms

Runtime Environment (OpenComRTOS)

Reference hardware platforms with 'Interacting Entities' architectural support

Runtime environment supporting network-centric concurrency and communication

Fig. 1.1 Unified Systems Engineering Methodology

computer models. Often modelling tools will be used which allow the designer to develop such a simulation model in a relatively short period of time. Simulation models are used to verify first of all that the requirements and specifications were correctly understood and analysed. They also allow to do quick "what-if?" analyses. Furthermore, this activity will result in improved requirements and specifications as it will contribute in further insight in how the system will behave under the fault and test conditions. Simulation models have another benefit: often they are a lot less expensive to develop than the real system and making a change has a lower cost than having to change a real prototype or working system.

If the project constraints allow it; one or more architectural models will be developed. In the ideal case and if the domain and the tools allow it, architectural models should be deduced automatically from the simulation model. After a trade-off analysis the best simulation model will become the implementation model. Architectural modelling will always benefit from being expressed in the highest possible level of abstraction, often domain independent, with the lower levels details being generated by a domain specific "compiler". It is assumed that this compiler is efficient enough for its output to be economically acceptable.

Finally, while a simulation and an architectural model allow us to reason and think about the system, they provide no proof that the system under development will be trustworthy. To verify this for critical properties (e.g safety related ones), formal techniques can be used. Formal models can be considered as a special case of simulation models. Often, formal techniques will result in fairly abstract models of parts of the system that can be fully automatically analysed and proved by formal model checkers. Such model checkers automatically traverse the whole state space and herein lie their strengths as well as their limitations. The formal model must be small and abstract enough to allow a complete model check in a reasonable amount of time (and memory resources). In the alternative, formal proving techniques can be used, but they require higher mathematical skills and are rather labour intensive. When used however, they can greatly increase the trustworthiness of the system.

Specifications are normally fulfilled by parts of the system being developed. For example, if the requirement of a radio receiver specifies a quality assessment for the signal in the presence of noise, the deduced specification will state a signal to noise ratio with a specific value expressed in dB and to be reached in a certain frequency band. One of the parts of the system that will be implemented to achieve these figures could be a filter component. In the methodology we call a subsystem component that realizes the specification an "entity", an abstract term for any part of the system. Developing such a subsystem entity might be a project on itself, for example a processor chip will be a component of a larger system but when the package is opened, it will show to be a complex system as well, albeit with much smaller physical features.

All subsystem entities together create the system because they interact. The system properties will be the result. Often, it makes sense to distinguish an interaction as a separate conceptual entity, especially if physical properties have to be taken into account. In general, entities and interactions can cross many different domains. For example, a light sensor will transform light energy from the environment into an electrical current. This sensor might have been mounted on a mechanical shaft so that it measures a position. The electrical signal can be used by a programmable controller that transforms the analogue signal into a digital one, calculates in software (or in hardware) a steering value that is transformed again into an analogue signal that is applied to an actuator that transforms the electrical signal into an hydraulic pressure. If one follows this chain, one can easily differentiate multiple entities and interactions of different types. At an abstract level however, we only see information, matter or energy being transformed and passed on between entities. If we would write a software simulator, it would be enough to have the equations or use the same message passing primitive. One might even replace the sensor by a software process running on the controller without changing the functionality of the rest of the system.

The latter is a crucial observation. Requirements and specifications express properties about the system without a reference to how they are realized. Entities and interactions express the architectural composition of the system. Expressing and analysing this is often called modelling. However, for this modelling to work smoothly, it is clear that it should be straightforward to move from one domain to another. Given that the real world is naturally expressed as sets of interactions and entities, it is straightforward to use the same paradigm everywhere, in particular for developing the architectural models and a selected implementation.

1.3 Interacting Entities for the Software Domain

Given that software programs are essentially computer models of real-world concurrent systems, and that real-world systems can often be described in terms of interacting entities, would it not be natural and beneficial that computer programs reflect this as well? Unfortunately, this is seldom the case. Most programs are written as a single long sequence of instructions, even if embedded systems technology allows for asynchronous interactions with the environment through an interrupt mechanism. One of the reasons for this lack of concurrency in the program is that the most widely used programming languages (C in the embedded case) have little or no support for concurrency. The fundamental reason is that processors evolved all from the sequential von Neuman machine architecture and that programming languages were developed bottom-up, from the hardware to the software and not the other way around as good systems engineering would have prescribed. The underlying issue for developing reliable systems is here again the state space. By keeping things sequential, the whole state space is exposed to the application layer and the larger the program, the higher the probability that a single error or fault can bring the whole system to a halt.

Nevertheless, in the embedded world, solutions were developed that addressed some of the issues. Concepts like multi-tasking and inter-task synchronization can be found in what we call RTOS (Real-Time Operating Systems). They were introduced as a means to decouple the timing behaviour from the functional behaviour in an embedded system, even at first sight at the cost of a little overhead. In embedded systems, real-time requirements are measured in microseconds and achieving predictable real-time properties while using big loops is very difficult, certainly in an implementation independent way. This must be seen in contrast with the desktop world, where the available computing power is much more abundant, but were real-time constraints are often very flexible and measured in terms of human perception. For vision this is around 40 ms but often even a few seconds can be acceptable.

Currently, several forces are converging to make concurrency in programming feasible as well as desirable.

1.3.1   Silicon Technology Advances

Contrary to may other domains, in electronics the rate of technological progress has been phenomenal. Engineering has succeeded more or less in quadrupling the amount of transistors on a given square millimeter of chip surface every two years. This means that higher clock rates, lower power consumption and functionality have evolved along similar lines. At the same time, this has made it rather straightforward to start putting multiple processors in a single device.

However, this has created a software productivity gap. Software is still largely being developed by the same human brains as 50 years ago using more or less the same (sequential) programming languages. Concurrent programming helps on several levels:

- Mastering the complexity: dividing the application in a number of smaller, more or less independent entities, breaks up the state space into smaller chunks as well.
- Greater reuse: when functions are programmed as concurrent tasks or processes with clearly defined interfaces (not through a shared state space but through protocol based message interaction), such software components can be reused with a plug-and-play technique even across multi-processor systems.
- Increased complexity of applications also means that larger software teams are needed. Again, concurrent tasks or processes can then be developed as components allowing to distribute the work.
- Increased need for reliability: as outlined above, the major issue for achieving reliability is the state space size. By confining the state spaces, we can limit the impact of an error to a smaller part of the system, especially in combination with hardware protection mechanisms in the programmable hardware.

1.3.2   Silicon Technology Limitations

The increased processor speeds do not fully result in an equal performance increase at the application level. Programs essentially transform data that is stored in memory and save it back to memory. However, memory technology does not scale as well as processing technology for size and access speed. Hence, when the data is in external memory, often the CPU will have to wait for hundreds of clock cycles. And while all kind of micro-architectural 'tricks' like caches, pipe-lining and other are in use to alleviate the performance gap, they do not address the core problem of real-time predictability and power consumption.

With increasing chip size, a bottleneck also appears on the chip itself. Most chips still use a single master clock and given the high density of the chips, it becomes increasingly difficult to distribute the clock signals across the whole chip. A natural solution is to define multiple concurrent clock domains. As it has become a lot simpler to implement fast communication channels between processor cores, even when going off-chip, a concurrent architecture emerges at the hardware level. As a result, depending on the computation to communication ratio of the applications as well as of the available hardware, it becomes feasible to achieve better application level performance using multiple slower processors rather than using a single very fast one. It saves on energy and it also saves on connection pins, often one of the most expensive parts of a packaged chip. The precondition is that applications are from the start designed in a way that they can easily be distributed over multiple processors. An additional benefit is that the system will consume less power and slower technologies, costing less can be used.

1.3.3   The World Becomes Connected

Advances in technology (in the software as well as the hardware domain) have also resulted in a major change in the way systems are deployed. The world became wired and internet allows to establish connections to almost any point in the world. In this world, a lot of existing technology has already been deployed and it is mandatory to connect with legacy systems in symbiosis. Again, a programming model that assumes concurrent and distributed platforms from the beginning makes this a lot easier to achieve.

As a conclusion, we can say that there are many reasons why a concurrent programming model is a natural one. It fits much better with engineering design activities like modelling and it fits best with the change in technology. This book describes a programming model that actually achieved this goal of universal (concurrent) programmability.

1.4   A Link with the Work Plan in a Systems Engineering Project

Finally, when all requirements, specifications and models have been developed we still need to plan to develop the product in a systematic way. The result is a work plan and typically the subject of "project management". Often, this work plan will start when specifications have been frozen and the design has been selected. Note that the activities we defined above as well as the development will not occur in a sequential order. A good engineering process should be iterative because every activity will allow discovering issues that were not considered before. This is also teamwork, with each team member providing a specific view on the system being developed. For this iterative process to take place, development should also be grouped into sufficiently small work packages with each task being small enough to keep an overview. Given the decomposition into entities, an entity or a group of related entities becomes a natural way of distributing the work to be done.

We define a Work Package as a group of related tasks that together achieve the implementation of a number of entities. We distinguish four main types of tasks, this distinction is important in the context of formalized development. The first type is the development task itself. Using specifications and the developed models, this type of task will develop the subsystems that as a whole result in the final system. If the modelling tools are powerful enough, then the implementation can often be generated. If not, a tedious work of manual coding may be needed. In order to verify that these development tasks were done correctly (e.g. as defined by a procedure) a verification task is defined. It will not verify the subsystem itself, but mostly verify that it was correctly developed. Once that is done, the developed entity can be tested in a test task according to the defined test cases together with the specifications. Finally, when all developments have been verified and the results tested and approved, everything can be integrated and validated against the requirements in a validation task. Again, the architectural paradigm of interacting entities will help here because it provides separation of concerns and they help to tackle the complexity.

1.5   System Engineering Methods and Engineering Standards

It is interesting to have a look at how professional engineers today achieve high reliability. Standards for system (and software) engineering exist and were developed to meet the requirements of developing complex (and hence costly) and often safety critical systems. Typical examples are found in the aerospace and transport sector. The standards were partly developed to achieve a more predictable and repeatable engineering process. Combine this with the need of certification by external third parties and most of these standards read like complex and confusing recipes.

If one reads these standards, one will notice that formal methods are mostly absent although when using them the developer gets "extra points". For the highest safety integrity levels they are a must in the sense that their use is required to prove for example that the software is correct under all circumstances. Most of these standards are also not very prescriptive but they list a large number of activities that should be taken when developing a safety critical system and they are used as a basis for certification. At this stage one must consider that such standards are first steps away from fully heuristic development towards fully formalised developments away from fully heuristic developments. Their use will certainly improve the resulting trust in the developed product or system but it is not a sufficient condition. We believe that as experience and education grows, formalisation of the system engineering process will increase followed by the development of cleaner and leaner process models. For example while IEC-61508 is considered the mother of safety engineering standards, it is almost impossible to apply with a lot of guesswork. On the other had standards like ISO-26262 and DO-178B are a lot more readable and define a more logical process. In the future we can also expect standards from different domains (e.g. aerospace, defense, railway, automotive) to merge. After all good systems engineering is not bound by a domain, although domain specific heuristics remain important as well.

1.6 Where Do Formal Techniques Fit in?

Traditionally, formal techniques (or formal methods) are seen as a technique to prove that e.g. a certain piece of critical software is correct. It must be understood that this is only partially possible because there are many dependencies. First of all, software is most often written in a programming language that was not formally defined and for which the compilation tools were not formally developed, neither is the processor developed and verified in a formal way. Hence, the programming language and the compilation tools are trusted on the basis of real-world use and extensive testing. A second issue is that the translation between the formal, often more abstract models and the programming language will never be "perfect". While formal techniques force us to specify and think about every fine grain level detail, at the same time they allow us to abstract away from the real world artefacts often introduced during a heuristic development process. Hence, one can prove that the formal model was correct, but that does not prove that the software as an implementation of the formal model is correct. Nevertheless, using these techniques can greatly improve the trustworthiness in the final system, partly because formal techniques force us (or rather allow us) to think more upfront about all possible states and ways to construct the system.

Considering the iterative systems engineering methodology described above, it is clear that formal techniques can be used in any stage of the process. The reason is that they help to reason about the system at a more abstract level with less interference from the heuristic knowledge that goes with implementation. This has proven to be a crucial factor in the formal development of OpenComRTOS. Because formal techniques were used from the very beginning to develop the architecture, several issues were detected that would not have been found after the implementation. It also allowed to develop a much more efficient and cleaner architecture, highlighting an aspect that is often neglected in systems and software engineering. An efficient architecture is the equivalent of a better algorithm. Whereas a better algorithm will often result in more efficiency for a specific property (performance, power consumption, etc.) a better architecture has often wider ramifications like being easier to use, easier to reuse and easier to adapt to other purposes. In this project, it even results in a much smaller code size (5–10 times smaller) than what previously had been obtained using traditional manual development. This is remarkable as an RTOS has always been perceived as a complex and difficult piece of software requiring the use of "hackers" to get it right. The formalized approach we took using formal methods did better without requiring extra resources and time.

Finally, it must also be said that one has to consider formal techniques as a crucial but not the only factor contributing to a trustworthy and optimized design.

After all, engineering is a human activity and there are many aspects and views that need to come together. Our experience showed that intensive team work is very beneficial, especially when combined with "formalisation". Formalisation serves two purposes. Firstly, it forces the participants to formulate their thoughts clearly while exposing the hidden assumptions. Secondly, the formulated concepts have to be made much more concrete and unambiguous. This allows to share the semantics which means that it becomes easier to discuss the system to be developed without guesswork filling in the gaps. Formalisation in itself will often be helpful in finding reasoning errors (typically finding unwritten assumptions) and the same procedure of formalization is also the necessary step toward the use of formal techniques. The latter can be considered as the next logical step in formalisation by translating the model in a format that makes it susceptible to mathematical analysis. This will in itself restrict the interpretation that can be given to the concepts used in the formal model but this is necessary to take control and to allow automated model checking.

Chapter 2
Requirements and Specifications for the OpenComRTOS Project

In this chapter, we discuss the requirements and specifications for the OpenCom-RTOS project from the point of view of its capabilities to support applications in meeting real-time requirements. As this is related to a distributed real-time operating system this is rather unique as most RTOS are designed for single processor systems and if not, they assume a shared memory architecture whereby the address space is global. OpenComRTOS on the other hand makes no such assumptions, but assumes that a processor has local memory and that the hardware allows to communicate somehow between the processors' memory. It is a network model, but allowing to emulate it by shared memory if this is available. Hence the term "network-centric" RTOS. In this context, we also explain the CSP background that is still present in the conceptual design of OpenComRTOS.

2.1 Background of OpenComRTOS

The initial purpose for developing OpenComRTOS was to provide a software run-time environment supporting a coherent and unified systems engineering method-ology based on "Interacting Entities". This was originally developed by Open License Society, (OLS 2011) and currently further developed and commercialised by Altreonic, (ALT 2010). In this methodology, requirements result in concrete specifications that are fulfilled in the architectural domain by concrete "entities" or sets of entities. Entities can be decomposed as well as grouped to fulfil the specifica-tions. In order to do so, we also need to define "interactions", basically to coordinate the entities. In practice these interactions can be seen as being protocols whereby the entities synchronise and exchange data, at least in the domain of embedded systems. The approach, however, is universal and the same view holds for other domains as well, e.g. business processes or mechanical systems. The difference is in the physical nature of the interactions and entities and in the terminology used. Interactions and entities are first of all abstractions used during the modelling phase. As such, a specified functionality can first be simulated as part of a simulation model,

E. Verhulst et al., *Formal Development of a Network-Centric RTOS: Software Engineering for Reliable Embedded Systems*, DOI 10.1007/978-1-4419-9736-4_2, © Springer Science+Business Media, LLC 2011

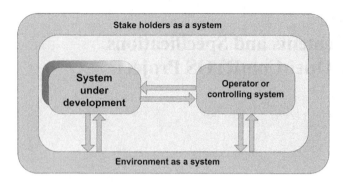

Fig. 2.1 The context of systems engineering

critical properties can be formally verified using formal techniques and finally an implementation architecture can be defined using the architectural modelling tools of the target domain. In our case, we try to keep the semantics unified from early requirements till implementation. In the targeted embedded systems domain, this means that the final architecture is likely a concurrent software program running on one or more programmable processors. Some functionality might be implemented in specific hardware entities. Such entities will be integrated in the input or output subsystem or will be designed as co-processing blocks. In most cases, these hardware entities will be controlled from software running on a processor.

In an embedded system, and in most systems, two additional systems must be taken into consideration, as illustrated in Fig. 2.1. The first one is the "environment" in which the embedded system is placed. This will often generate inputs to the system or accept outputs from it or it will influence the operating conditions, not necessarily in a fully predictable way. A second system that is often present is the "operator", who also will generate inputs or act upon the outputs. If this is a human operator, we have to deal with an entity who's behaviour is not necessarily always predictable. Often, the "operator" might be another embedded system and then the behaviour should be more predictable, at least if well specified. However, systems are layered. If we "open" the embedded system or consider the system under development with its environment and its operator as a new system, we can see that each system can be a component in a larger system and often it will be composed itself of "subsystem components". For this book, we stay at the level where such components are programmable processors.

The use of a concurrent programming paradigm embodied in an RTOS is then a natural consequence of the unified semantics paradigm. Programming in a concurrent way implies that the abstract entities (that fulfil specifications) are mapped onto RTOS "Tasks" (also called processes or threads in the literature) and that interactions are mapped onto services used by the tasks to synchronize and to communicate. In principle, this abstract model equally well maps to hardware as to software but we focus here on the software. The target domain ranges from small single chip micro-controllers over multi-core CPUs to widely distributed heterogeneous systems that include support for legacy technology. OpenComRTOS should allow to program in a transparent way such a target system, independently of the processor or communication medium used. In the context of the OpenComRTOS project, an additional requirement is related to the process used for developing OpenComRTOS. Given the importance and generic nature of a RTOS as runtime layer, we considered it important that the resulting RTOS would be safe, correct and performing. Hence, it was a requirement to use formal techniques for its design as well as for its verification.

2.2 Early Requirements Derived from the Virtuoso RTOS

Following the requirements, we can in principle derive the specifications. However, specifications cannot be fully separated from implementation choices. In our case, a previously developed RTOS called Virtuoso (Verhulst 1993a,b, 1997a,b, 2002) served as a guideline.

In this section we first analyse it, highlighting issues as well as benefits of the then existing architecture. Virtuoso was a distributed RTOS, developed by Eonic Systems until the technology was sold to Wind River Systems in 2001. Its main target was parallel DSP systems, although it has been ported to other architectures in single as well as in multiprocessor versions. Even heterogeneous target systems have been demonstrated albeit at the expense of a lot of manual integration work. The smallest system it had been ported to was an 8bit micro-controller in a telephone handset, the largest system was a system with over 12,000 processing nodes, heterogeneous but built up as a number of homogeneous clusters. The experience with Virtuoso was crucial. It also taught us the limits of such a programming environment. These limits lay not so much in the RTOS itself but in maintaining and supporting it. As such the porting effort was high because of the complex and optimised architecture and it was very difficult to add functionality. Nevertheless, its overall functionality of transparent parallel processing (this was called the "Virtual Single Processor" runtime model) (Verhulst and De Jong 2007; Verhulst et al. 2008) was a major driving force to redevelop it in a better way. Hence, to some extent, OpenComRTOS is conceptually a fourth generation of Virtuoso although it was a clean room development. At the outset we were curious to see how we could do better but it was totally unanticipated we would be able to improve it beyond all initial expectations.

The Virtuoso RTOS had its origin in the pioneering INMOS transputer (Wikipedia 2011; INM 2011), a partial hardware implementation of Hoare's Communicating Sequential Processes (CSP) process algebra (Hoare 1985a) masterminded by Dr. David May. Later on Virtuoso was ported to traditional processors but mostly parallel DSPs. The transputer was a rather unusual RISC like processor with unique support for on-chip concurrency and inter-processor communication. On-chip it had a scheduler with two priority levels, each level supporting round-robin scheduling between the compile time generated processes. It also had hardware support for inter-process communication and synchronization using "channels". For distributed, embedded real-time applications, it raised two major issues:

- Two levels of priority are not enough for hard real-time applications. Typically, at least 32 levels of priority are needed and full support for pre-emption and priority inheritance.
- Topology independence: although the transputer had interprocessor links, the communication between processors had to be manually routed at the application level. The issue is here mostly one of maintenance. Every little change in the topology could result in major reprogramming efforts.

Although the transputer hardware and available software (like occam and parallel C) were derived from the CSP process algebra, the use of processes and channels was very novel and hence this was a stumbling block for wide adoption by the embedded market. That and the failure to develop a successor product (there was one but it was over-specified for the silicon technology available at the time) resulted in the demise of INMOS, the company developing it. Nevertheless, it lives on as a set-top box controller from ST Microelectronics and the link technology is now adopted as a standard by the space sector (IEE1355 enhanced with LVDS signalling and subsequently called SpaceWire), see (Spa 2011; IEE 2011). One of the closest successor chips today is the XMOS chip, see (XMO 2011), also a brainchild of David May. While the CPU is a more convential RISC processor, it has support for hardware assisted multi-threading (up to 8 threads/core) and a single chip contains from 2 to 4 CPU cores.

The above observations resulted in the adoption in the Virtuoso RTOS of the following architectural principles:

- Use of 255 levels of priority with full pre-emption capability.
- Development of traditional RTOS services like events, semaphores, fifos, mailboxes, resource and memory maps.
- System wide identifiers used for transparent system wide routing.
- Use of command and data packets to provide topology independent programming.
- Packets carrying a priority inherited from the generating task.
- Support for priority inheritance in the scheduler (see further).

When Virtuoso was ported for the first time to a parallel DSP target, it was found that a major redesign was needed. While the links on the transputer could be accessed by memory mapping the internal channels to an I/O address, DSPs provide the programmer with a series of "raw" I/O ports. The challenge here is that the communication on these ports is interrupt driven, often with DMA engines operating in the background. On the Texas Instruments TMS320C40 DSP or Analog Devices 21060 DSP for example, this resulted in 12 interrupts and 2 timer interrupts that were independently enabled. Hence, there was a need to develop a low latency driver architecture. This was achieved by developing a specific system level nanokernel that was mainly written in assembler and used a subset of the processor context. While it provided very fast communication handling, it had the drawback that porting it to another processor was a major effort.

Later on, Virtuoso was also extended with new features and services. However, these were at the expense of performance while the code size was rapidly growing. Another observation was that this resulted in some errors that could remain dormant for years until an unanticipated usage triggered them.

Combining all these experiences resulted in some decisions at the architectural level that are found back in the OpenComRTOS requirements:

- General use of packets at all levels
- Prioritisation at all levels (local scheduling as well as at the communication layer)
- Support for distributed priority inheritance
- Support for traditional RTOS services (events, semaphores, fifos, etc.)
- General use of ANSI-C, minimising the use of assembler (hence no nanokernel)
- Modular architecture allowing to remove or add functionality without affecting the rest of the system
- All services to be formally modelled and verified

At the start of the OpenComRTOS project, these were not firm requirements. As the reader will see, the use of formal techniques in the project has allowed it to go way beyond the functionality of Virtuoso. Although it was known as highly performant and small in code size, we were not expecting to be able to do much better, except in having a cleaner and safer RTOS.

2.3   Real-Time Embedded Programming

While most programming is concerned with performance (often expressed in terms of achievable throughput), real-time is then often equated to "fast enough". In the embedded domain, however, the system will often interact with the physical world whereby stringent time requirements must be met or the system can fail. In such systems, the reactive behaviour is most important and must always be achieved independently of the logical correctness of the application. Such systems are often called "hard" real-time in contrast with "soft" real-time systems where time properties are statistical.

2.3.1   Why Real-Time?

It can be argued that an architectural paradigm based on entities and interactions does not need any notion of real-time. Indeed, the time properties can be considered as mostly orthogonal to the "logical" behaviour of a system. In the embedded domain (and most of the systems we use have embedded aspects), we are dealing with real-world interaction and time is part of it. Signals that the embedded system must process arrive in real-time and must be dealt with before the next set of signals arrives. Similarly, the embedded system will act on its surroundings and real-time requirements apply. Implicitly, we assume here that sampling theory is applied. Sampling theory dictates that we should at least sample at twice the bandwidth of the signal. Similarly, when we apply output or control signals this must also be done with a rate at least equal to twice the bandwidth. If the controlled subsystem has a mechanical mass and physical properties such that inertia determines the dynamic behaviour, we must similarly take into account its time constant. Sometimes, the output timing can be rather demanding. An example is audio processing. The human ear is very sensitive to phase-shifts so that even when the bandwidth requirements are met, the jitter requirements are stringent enough that hardware support might be needed.

The purpose of an RTOS is to give the engineer the means to meet such real-time requirements at the same time as he is meeting the architectural ones (as explained before: mapping abstract entities into concrete tasks). Timely behaviour is then a property of the tasks in a specific execution context. This allows to design and verify a real-time system without having to look into the details of the algorithms executed by the tasks. The only information needed is what resources the tasks use (e.g. time in the form of processing cycles and memory). Executing the task on another processor does not change the algorithm, just the timing and memory used. Similarly, a concurrent program in itself does not need to be real-time (it is a matter of defining the parameters differently). However, it is very convenient that a concurrent program that was designed to handle real-time, can also handle time-independent programming, e.g. for simulation purposes. The opposite is often not true.

2.3.2 Why a Simple Loop Is Often not Enough

It is useful for the remainder of this book to present in short our view on embedded real-time programming. The reader can find a wide range of literature related to real-time and embedded programming elsewhere if he wants to investigate in more depth.

Let us start with the term "real-time". The intuitive notion of real-time is often a subjective one using terms like "fast" or "fast enough". Such systems can often be considered as "soft" real-time, because the real-time criteria are not clearly defined and are often statistical. However, when the system that must be controlled is physical, often the deadlines will be absolute. An example of a soft real-time system is a video system. The processing rate is determined by the frame rate, often a minimum of 25 Hz and determined by the minimum rate needed for the eye to consider the frames as a continuous image. The human eye will itself filter out late arriving frames and can even tolerate a missing frame. Even more soft real-time are on-line transaction systems. Users expect them to respond within e.g. one second, but accepts that occasionally it takes tens of seconds. Of course, if soft real-time repeatedly violate the expected real-time properties the Quality of Service will suffer and at some point that will be considered a failure as well.

Hard real-time systems on the other hand that miss deadlines can cause physical damage or worse result in deadly consequences if the application is safety critical, even when a "fail-safe" mode has been designed in. Typical examples are dynamic positioning systems, machine control, drive-by-wire and fly-by-wire systems. In these cases often the term "hard real-time" is used to differentiate them from the former. From the point of view of the requirements, hard real-time means "predictable" and "guaranteed" and a single deadline miss is considered a failure.

Two conclusions can be drawn. First of all, a hard real-time system can provide "soft" real-time behaviour, but the opposite is not true. Secondly, when safety critical, a hard real-time system must remain predictable even in the presence of faults. In the worst case it could fail, but the probability of this happening must be low enough to be considered an acceptable risk.

Strictly speaking, no RTOS is needed to achieve real-time behaviour in an embedded system. It all depends on the complexity of the application and on the additional requirements. For example, if the system only has to periodically read samples from a sensor, do some processing and transmit the processed values, a simple loop that is executed forever will be sufficient. Sources of complexity are for example:

- The need to put the processor to sleep in between processing to conserve energy
- Managing several hundreds of sensors
- Executing at the same time a high number of other tasks
- Detecting a failure in the sensor circuit
- Detecting a failure of the processor

Such requirements are difficult if not impossible to handle when a simple polling loop is used, but as most processors will have support for interrupt handling, the developer can separate the I/O from the processing. This essentially means that most embedded systems will have a "hardware" level of priorities and a "software" level of priorities. The highest priority level is provided by the Interrupt Service Routines that effectively interrupts the lower priority (background) loop. However, the extra functionalities listed above might already require multiple interrupts and priorities. The sleep mode of the processor requires that the circuit generates an interrupt to wake up the processor and a timer supporting a time-out mechanism might be needed for detecting a failure. Also the transmission of the processed values might require some interrupts. Hence, the question arises how each interrupt must be prioritised. In the simple example given, this is not much of an issue as long as we assume that the system is periodic and always has spare time between samples. What happens however if multiple interrupt sources are present and if they can be triggered at any moment in time, even simultaneously?

2.3.3   Superloops and Static Scheduling

When multiple interrupt sources are present, a simple solution is to distribute interrupt handling and processing over the available interrupt service routines and the main polling loop. The separation between "handling" and "processing" of

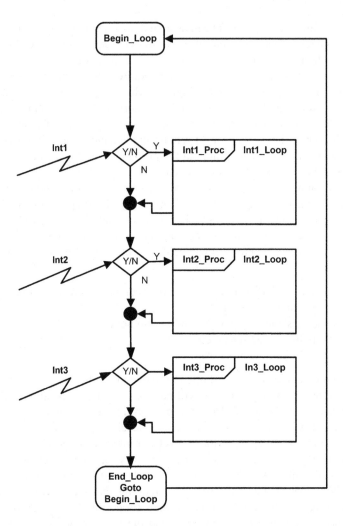

Fig. 2.2 Superloop schedule with three interrupt sources

interrupts is essential because interrupts will be disabled when an Interrupt Service Routine is entered and worse, the hardware might be designed in such a way that the data is only available for a short period of time. Hence, while an interrupt is being handled, the hardware must have a mechanism for holding arriving interrupts, else they will be lost and in the worst case, the application will fail. Therefore, interrupt handling should be kept as short as possible. On the other hand in the polling loop, the program will repeatedly test for the presence of the interrupt and when enabled execute the corresponding processing function (Fig. 2.2).

The issue is that such testing and processing must be done in sequence and that the program cannot progress unless the interrupt has arrived. Hence, if all interrupts are to be seen and processed, a static schedule must be calculated and the peripheral hardware must be configured to be compatible with it. Such a schedule is not necessarily feasible, e.g. when the arrival rates of the interrupts have a wide span and do not follow a harmonic periodicity. In addition, the polling will waste processing cycles that could be used for useful processing and worst, if for some reason the interrupt does not arrive, the whole system can block. From a safety point of view, such a polling loop has no built-in graceful degradation. In addition, even when no errors occur, a small change in the application can result in the need to recalculate the whole schedule or in the worst case can result in the application no longer being scheduleable. What we need is a separation of concerns. The logic of processing should be made independent of time. With a sequential loop (on a sequential processor), this is not possible because the state space is shared amongst all processing functions and in addition the time behaviour depends on the time behaviour of the rest of the processing functions. What is needed is a mechanism that divides the global state space into local state spaces.

There are two ways to achieve this:

- To dedicate a processor to each "local" processing function
- To create a mechanism that separates the state spaces, even when executed on the same processor

The first solution has as side-effect that interprocessor communication can now become an issue (because communication media are also shared resources). The second solution creates the concept of "virtualisation", in essence a mechanism whereby each local processing function has virtually access to the full state of the processor. Note that this is only really possible because time is allocated to each virtual state space and this essentially means that to meet the real-time requirements at system level, this allocation of time must be carefully done to meet all real-time constraints.

The two solutions introduce both the notion of "concurrency", whether physical or virtual. Most real-time applications will however have "interactions" (e.g. passing data or synchronisation of a state that was reached) between the local state spaces. In line with the need for separation of concerns, we need a mechanism that "virtualises" these interactions independently on whether they take place on different processors or on the same processor.

And last but not least, while we separated the time behaviour from the logical behaviour, hard real-time systems still need a mechanism for handling time. This mechanism is called scheduling. We have seen a static version of it at the beginning of this chapter, called static scheduling. It assumes perfect knowledge about the system when it is built and assumes that the system's operating parameters are static and will never change. As outlined, this is not always the case, certainly when failure conditions are taken into account. In general, a more dynamic scheduling mechanism is preferred. The scheduling can be based on a measurement of time or on the time already used. The most widely used mechanism is based on priorities, a ranking of the processing functions based on an analysis that combines the periodicity and the relative processing load. This mechanism is called Rate Monotonic Scheduling (RMS). OpenComRTOS is a RTOS based on the assumption

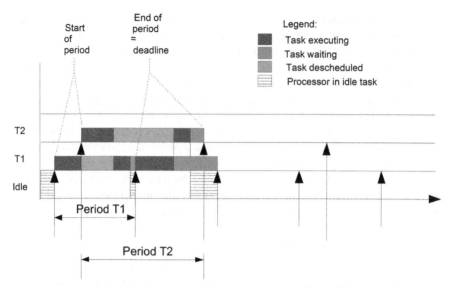

Fig. 2.3 Two periodic tasks scheduled with RMA

that a Rate Monotonic Analysis (RMA) is executed, resulting in a priority ranking of the scheduled application functions. Nevertheless, the design allows for the implementation of different scheduling policies which is likely to happen in future versions.

2.3.4　Rate Monotonic Analysis

RMA was first put forward in 1973 by Liu and Layland (1973). Although it does not solve all issues it provides a good framework that is simple and most of the time it is applicable. The algorithm states that given N tasks with a fixed workload that must be active with a fixed periodicity (with the beginning of the next period being considered as the deadline for the previous period), all deadlines will be met if the total processor workload remains below a value of about 69% and a preemptive scheduler is used with each task receiving a priority that is higher if the task has a higher periodicity. The upper bound of 69% is obtained for an infinite number of tasks. For a finite number of tasks and especially when the periods are harmonic, the upper bound can be a lot higher, often even observed to be above 95%. Figure 2.3 illustrates RMA scheduling of two tasks. In general, the RMA scheduling algorithms is as follows:

$$\sum_{j=1}^{n}\left(\frac{C_j}{T_j}\right) \leq U(n) = n \cdot \left(2^{\frac{1}{n}} - 1\right) \tag{2.1}$$

with:

- C_j being the worst case execution time of $Task_j$.
- T_j being the execution time of $Task_j$.
- $U(n)$ being the worst case utilisation with n Tasks.

According to equation (2.1) a system with one task has a utilisation of 1.0 ($U(1) = 1.0$). For an unlimited number of tasks the utilisation converges at 0.69 ($U(\infty) = 0.69$).

In practice, the results of the first RMA algorithm are a (pessimistic) approximation and rely on some assumptions that are seldom met in real applications. For example, all tasks are assumed to be independent (hence they all are activated on independent events and do not synchronise or communicate with other tasks, nor do they share any resources). Also, task activation is assumed to be instantaneous and the processor provides a fixed processing power (hence no cache effects). Even if often the 70% level is used as a maximum load in any case, this means that to remain on the safe side, it is often better to keep the overall CPU load lower than the figure obtained. On the other hand, if only a few tasks are used and the interactions are limited, often the application will miss no deadline even if the processing load is higher than 70%. The CPU load can also be higher if the periodicity of the tasks is harmonic. Hence, RMA has to be seen as a guideline that must be complimented with a detailed analysis, profiling and especially measures to give the application more margin. It also should be pointed out that if a RMA schedule misses deadlines for the lower priority tasks that the higher priority tasks can meet their deadlines. This property of preemptive priority based scheduling is e.g. useful for creating a highest priority task that is only activated when exceptions have to be handled.

A very detailed and comprehensive analysis of RMA is given in Briand and Roy (1999). It also discusses the follow-up RMA algorithms that were developed later on and taking into account realities like blocking times (using shared resources), inter-task dependencies and distributed systems. In all cases this does result in higher boundaries for the CPU workload. The most important change to the basic RMA algorithm is that for determining the task priorities, one should not use the full period but the pseudo period that is derived by taking into account that the deadline of a task happens often before its period has expired. This is called Deadline Monotonic Analysis (DMA). More extensive descriptions as well as algorithms for schedulability analysis for a wide range of RMA scheduling policies can be found in (Klein et al. 1993).

It must be said, however, that for distributed systems no real RMA algorithm exists, although tools like MAST (MAS 2011; Harbour et al. 2002), allow to verify that a given schedule is feasible. In practice, a good system design with priorities will give good assurances that all deadlines can be met.

An important observation is also that a rigorous and static design might not always give the safest system if the first missed deadline results in catastrophic behaviour. In practice, many systems can tolerate missed deadlines if these misses have a low probability and if they are spread in time (not bursty). Of course, this means that the system design must takes this into account. A classical example is a brake-by-wire system. It must be designed for the maximum speed of the car and hence often the maximum rate will be used all the time. Even at this highest rate, there will be margin as the time constant of the mechanical system will be lower. If the car then operates at a lower speed, the control rate can be lowered as well and missing control signals from time to time (but not in continuous bursts) will in the worst case only lower the "quality" of braking, but this is often not catastrophic.

2.3.5   Priority based Scheduling in OpenComRTOS

In OpenComRTOS, it was decided to support priority based preemptive scheduling as the standard scheduling policy. In (Briand and Roy 1999), this is called Highest Priority First. Every Task can be assigned its own priority based on an off-line Deadline Monotonic Analysis (DMA). It must be said however that DMA assumes that all tasks execute on a single processor, whereas OpenComRTOS supports multi-processor systems. Hence, priorities are considered as a system-wide scheduling parameter and the DMA should still hold locally on each processor.

OpenComRTOS was also designed to clearly separate Interrupt handling (in ISRs) and interrupt processing (in a task). Good design practice dictates that a minimum time is spent in interrupt handling to improve the responsiveness of the system and hence, because interprocessor communication often requires fast interrupt handling, it will reduce the latencies. The latter is especially important for multiprocessor systems as the processing can be distributed over several processors and the scheduling delay includes communication delays. Similarly, in the design of a network-centric RTOS it was recognised that delays can also be the result from implementation artefacts. Hence, any activity in the RTOS or systems level drivers is done in order of priority. This minimises the point-to-point latency. Typical cases where this can be important are waiting lists and interprocessor communication. This means that one should be able to ignore the different scheduling latencies as the communication delay can be more important (especially on slow–speed networks). This latency is a combination of several factors that are difficult to quantify. Factors are: communication load, communication driver set-up time, transmission delay and receiver latency. Therefore, good profiling tools are a necessity. DMA then provides a good approximation and starting point. For extreme processor loads (typically, when the task's individual processing time is of the same order of magnitude as the system latencies), this assumption does not hold and often only static scheduling or dedicating processors to such loads is the only acceptable solution.

A small note however on the assignment of the priorities. In our case, these are assigned at design time and the scheduler is a straightforward Highest Priority First one. Research on dynamic priority assignment (Styenko 1985) have shown that algorithms that use Earliest Deadline First (EDF) algorithms (the priority becomes higher for the tasks whose deadline is the nearest) can tolerate a workload of up to 100%. There are, however, three reasons why this option was not further considered. The first one is that the implementation of an EDF scheduler is not trivial because measuring how far a Task is from its deadline requires that the hardware supports measuring this. As this is often not the case, one has to fall back on software based solutions that periodically record the task's progress. For reasons of software overhead, this must be done with a reasonable frequency, typically about 1 ms which means that fine-grain microsecond EDF is not feasible (1 ms can be quite long for a lot of embedded applications). The second reason is that no algorithms are known that allow to calculate the EDF schedule on a distributed target. The third, but fundamental reason is that an EDF schedule has no graceful degradation. If a task continues beyond its deadline, it can bring the whole system down by starvation, whereas a static priority scheme will still allow the highest priority task to run. This task can be activated by a time-out mechanism so that it can terminate such a run-away task before the other, still well behaving tasks are starved. Hence, if EDF scheduling is used, it is better to restrict this to a maximum priority level within a standard priority based scheduling scheme. A similar observation will be made in the next section when discussing priority inheritance schemes.

A general remark must be made here. A RTOS in itself does not guarantee that all real-time requirements will be met. Designers must use scheduleability analysis tools and other analyses like simulation and profiling to verify this before the application tasks are executed. However, an RTOS must provide the right support for executing the selected schedule. In general, this means a consequent scheduling policy based on priorities with pre-emption capability and with support for priority inheritance. OpenComRTOS provides this complemented with a runtime tracing function allowing to profile the time behaviour at runtime.

2.3.6 The Issue of Priority Inversion and Its Inadequate Solution

A major issue that has a serious impact on predictability is the presence of shared resources in an embedded system. A shared resource is often associated with a critical section or an access protocol. The latter are needed to assure that only one task at a time can modify the status of the shared resource. Examples are:

- A shared memory buffer that must be read out before new data is written in.
- Hardware status registers that set a peripheral in a specific state.
- A peripheral that can handle only one request at a time.

Note that a shared resource is a concept at a higher level of abstraction than the physical level but it will often be associated with it. It can be used to protect a critical section (e.g. the update of pointers in a datastructure) but it is not a critical section in itself. The critical section is a sequence of steps of the updating algorithm that must be done in an atomic way to guarantee that the datastructures remain coherent. It should also not be confused with disabling interrupts on a processor. The latter is a hardware mechanism that is processor specific and is designed to prevent other external interrupts from interfering with the intended program sequence.

In the context of a concurrent program, resource locking means that the system assigns temporarily ownership of the resource to a specific task until the task releases the resource. If more than one task requests to use the same resource, the second and subsequent requesting tasks cannot continue and will be blocked until the resource is released by its current owner. During the time a task owns a resource, it can get descheduled, e.g. because another higher priority task becomes active, the task requests a second resource, the peripheral associated with the resource is delayed itself or the task needs to synchronise with another task that has lower priority. In all cases, the resource owning tasks and other waiting tasks can be blocked from progressing which means that deadline violations become possible even if the priorities were correctly assigned and the application is scheduleable with known blocking times.

A very important conclusion to draw at this point is that a good design will try to limit the blocking times as much as possible and should avoid needing to protect shared resources at all. This might require a change in the architecture of the system but from the reliability and safety point of view this is a cheap preventive measure.

The real issue comes in when we also analyse what can happen as a function of the assigned priorities. Assume a high priority tasks request a resource that is owned by a low priority task. As it is a low priority task, middle priority tasks that are ready to run will pre-empt the lower priority task and if they have lengthy processing times, they can block the high priority task even if they do not need the resource at all. This problem is called priority inversion and was made famous in 1997 when the Mars Pathfinder kept resetting itself as a result of a continuously missed deadline, which was caused by a classical case of priority inversion as described above.

Is there a cure for this problem assuming that the system architect did his best in minimising the need for resource locking? The answer is unfortunately no, but the symptoms can be relieved. The solution is actually very simple. When the system detects that a task with a higher priority than the one currently owning the resource is requesting it, it temporarily boosts the priority of the current owner task, so that it can proceed further. Priority inversion will be avoided. In practice different algorithms were tried out, but in general the only change made is that the boosting of the priority is limited to a certain system specific ceiling priority. Else, the scheduling order of other tasks requiring a different set of resources can be affected as well and it might prevent that the system as whole achieves its goal.

If we analyse the issue of blocking in the context of a real system, we can see however that the priority inheritance algorithm does not fully solve the blocking issue, it relieves the symptoms by reducing the blocking times but a good design can maybe avoid them in the first place. During this project we found also that the resource blocking issue is part of a more general issue. In essence, a concurrent real-time system is full of implicit resource requests. For example, if a high priority task is waiting to synchronise with a lower priority task, should the kernel not also boost its priority? To make it worse, if such a task is further dependent on other tasks and we would boost the priority can this not result in a snowball effect whereby task priorities are boosted for all tasks and of course, we would have no gain. Or assume that the task is waiting for a memory block while a lower priority task owns such a memory block. Or assume that a task acquires a resource, which makes it ready and is put on the ready list. But while it waits to be scheduled a higher priority task becomes ready first and requests the same resource. Which means that the first task that was ready should be descheduled again and the resource given to the higher priority one (Fig. 2.4).

While all these observations are correct, often such situations can be contained by a good architectural design. The major issue is that implementing this extra resource management functionality is not for free and the tests they require are executed every time, resulting in a not-negligible overhead. The conclusion is that in practice resource based protection must be avoided by design and that priority inheritance support is best limited to the traditional blocking situations. In the case of the implicit resource blocks, if they pose an issue to the application, they can be reduced to a classical priority inversion problem by associating a resource with the implicit resource. For example, if a memory block is critical, associate a resource at the application level and normal support for priority inheritance will limit the blocking time. Else make sure that the system has additional memory blocks available from the beginning.

2.4 Next Generation Requirements

In the previous part of the chapter, we have limited ourselves to the handling of real-time requirements. An unspoken assumption was that the system is fully defined at compile time. For most embedded applications this is the case. However, as applications are becoming more dynamic and adaptive, the complexity is increasing as well. In such applications, meeting stringent real-time requirements is still often a prime requirement but it is not sufficient. The real-time requirements will have to be met when running multiple applications simultaneously with a variable amount of available resources. In the extreme, this also means in the presence of faults resulting in a number of resources no longer being available on a permanent or temporary basis.

We will illustrate this with two use cases for which the network-centric Open-ComRTOS could provide the system level software.

The first use case is a next generation electric car. Such a car will be fully controlled by software and electronic components ("drive-by-wire") and likely have a distributed power and wheel control architecture whereby for each wheel power control is combined with active suspension control, stability, anti-slip control and even braking. Many components can fail or show intermittent failures, e.g. sensors can fail, wires can break, connectors can give micro-cuts (very short absence of electrical contact due to vibrations), memory can become corrupted, processors can fail, etc. While the design should be robust enough to make such failures very low probability events, over the lifetime of the car such occurrences are certain. Practically speaking, this means that while the system can be designed assuming that all resources are always available, the designer must provide additional operating modes that take into account that some resources are not available for meeting all requirements. In the simplest case, this can mean that when one wheel controller

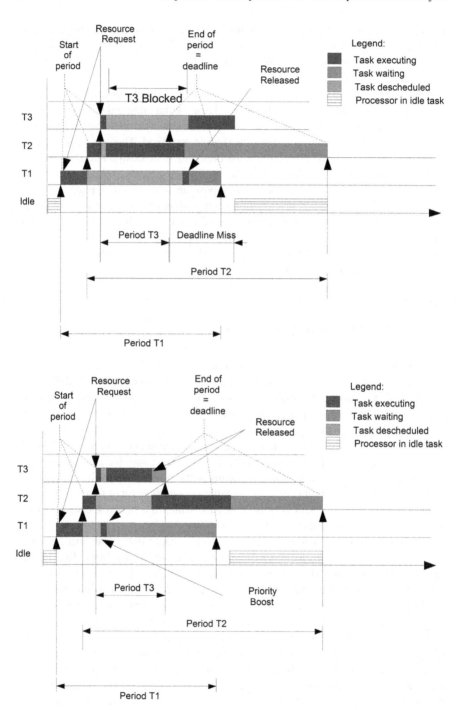

Fig. 2.4 Three tasks sharing a resource with and without priority inheritance support

fails, the processing is immediately redistributed over the three still fully functional units. Or this can mean that the system switches to a degraded mode of operation with a different set of tasks using less compute intensive algorithms.

The second use case is a next generation mobile platform. It is envisioned that such a platform will have tens of processing nodes, execute multiple applications with some applications showing a variable processing load depending on the data being processed (typical for multimedia and image processing). In the worst case, the processing load can even surpass temporarily the available processing power. On the other hand such applications can often tolerate a few missed deadlines. However, such a mobile platform loaded with a dynamic set of tasks, poses additional constraints. For example, when using wireless connections, bandwidth will vary over time, processing power might be variable because of voltage and frequency scaling techniques to minimise power consumption and available memory will vary depending on the use by other applications.

What these two use cases illustrate is that an embedded real-time application is becoming more challenging for following reasons:

- All applications can no longer be defined fully statically.
- Some applications have a variable processing load.
- The system software must not only schedule the use of time as a resource, but also other system resources like bandwidth, processing power, memory and even power usage.
- The system will have hard real-time constraints as well as soft real-time constraints.
- The system will have different "modes" (each consisting of a coherent set of states).
- Fault tolerance is not to be considered as an exception but as a case where the system has less resources available.

The result is that such an embedded system becomes "layered" and time as a resource is not the only one that must be scheduled. Such a system will need to schedule the use of several resources. In the guaranteed mode of operation, we find back the traditional static scheduling. Rate Monotonic Scheduling provides for meeting the time properties whereas compile-time analysis assures that all other resources are available. In the extreme case, this includes providing for fault tolerance because the system has to be designed with enough redundant resources to cope with major failures. The next layer is then a best-effort mode in which the properties are guaranteed most of the time, eventually with degraded service levels. For the time properties this means we enter the domain of soft real-time, but often at the application level this means the system offers a statistically defined level of quality of service level. A typical example is generating an image with less resolution because not enough processing power was available during the frame time. In the extreme case this corresponds with a fail-safe mode of operation whereby the quality of services is reduced to a minimum level that is still sufficient to stop the system in a safe way. Finally, the last layer is one where essentially nothing is guaranteed. The system will make only resources available if there are any left. Statistically, this can still be most of the time unless a critical resource like power is starting to fail, and the system then was designed to put the processor in a "sleep" mode to e.g. stretch battery time.

What we witness here is a transition from a statically defined hard real-time system with fully predictable time behaviour, but possibly catastrophically failing, towards a system where the design goal is defined as a statistical quality of service (QoS) at the application level. Such a system must still be able to meet hard real-time constraints in a predictable way but must also offer different operating modes corresponding with a graceful degradation of the services offered by the system as a whole. Practically speaking, when a processor fails, it will often be catastrophically although processors with a MMU (Memory Management Unit) and appropriate system software can contain the failure to the erroneous process without affecting the rest of the processes. Most embedded processors however will need a hard reset to recover from such a fault. Hence, such a system will need redundancy of hardware resources, be it as part of a distributed system, be it as part of a multicore chip.

These next generation requirements were not addressed in the OpenComRTOS project discussed in this book, but the fact that OpenComRTOS supports programming a multicore and distributed system in a transparent way facilitates addressing such requirements.

2.5 Top Level Requirements for OpenComRTOS

OpenComRTOS was developed from scratch as an answer to the need to have a uniform programming environment fitting within the notions of "unified semantics" and "interacting entities" we put forward in the introduction as key paradigms for a systematic systems and software engineering method. Furthermore, we restricted the application domain to embedded systems, often distributed and operating within strict boundaries of predictable real-time behaviour and resource constraints.

If we analyse these top level requirements, we can define following top level properties as requirements for OpenComRTOS. Note however that this is not the complete list of requirements.

- *Scalability*: OpenComRTOS shall support the redeployment of applications, mostly by recompilation of the application source code, from very small single micro-controller systems to target systems with a large number of distributed heterogeneous processing nodes. As a result scalability works in two directions. An application can be remapped on more or less processor resources without the need to modify the application itself. This capability should be independent of the underlying processor architecture (from 8 bit to 64 bit CPUs).
- *Real-time support*: OpenComRTOS shall allow to develop applications that are predictable in time, including when multiple processors are used. A real-time capable scheduler with support for priority inheritance shall be included.

- *RTOS conformant API*: The OpenComRTOS API shall reflect commonly used services as available with existing RTOS like task management, events, semaphores, fifos, resource locks, etc.
- *Portability*: OpenComRTOS should be straightforward to port to another target system. A precondition is the use of ANSI-C. The C-code shall have no defects and compile with no warnings.
- *Extensible*: OpenComRTOS can be extended with application specific services and entities without the need for the user to develop another middleware layer or without the need to redevelop the RTOS kernel. Such services are integrated at the system level.
- *Distributed operation*: Whether the application tasks and the kernel entities are placed on a single processing nodes or are mapped onto several ones, the developer does not need to care about where the tasks and entities are mapped (except at configuration time and when considering performance issues). The system itself takes care of the routing and system level communication. As a result, the application source code is independent of the target's topology. Note that a special case of target networks are multi-core chips whereby the CPU cores are linked with a dedicated NoC (Network-on-chip). Shared memory is allowed if the right protection mechanisms are used.
- *Efficiency*: Efficiency can be measured in terms of functionality for a given code size. Small code size means that less time is spend in executing kernel services resulting in a lower overhead. As OpenComRTOS is network-centric, a major measure of efficiency is the latency to set up a communication or to start a kernel service.
- *Safety*: On traditional processors it is very hard to provide safety unless the processor has dedicated hardware support, the latter mainly being available on high-end processors. The main source or runtime errors is linked with a lack of available memory and buffer overflow. The architecture should be developed in such a way to minimise such risks and allowing the use of hardware support when available.
- *Runtime profiling*: Support shall be provided for examining the kernel and application task state at runtime as well as for profiling execution traces.
- *Formally verified*: In order to minimise the risk of remaining errors in Open-ComRTOS, the architectural design and the implementation shall be modelled and verified using formal techniques.
- *Automation*: As many programming errors are due to manual coding by software engineers, OpenComRTOS must be supported by a graphical development environment and other supporting tools providing a higher level of abstraction and generating the application code as much as possible automatically from a higher level model description.

2.6   Specifications Derived from Requirements

From the rather informal requirement statements, we derive verifiable specifications. The various terms used will become clear in further chapters.

- *Architecture*

  - The RTOS shall use Packets for service invocation at all levels.
  - The Packet structure shall be identical for all processors.
  - Packets shall contain a header and a payload section.
  - The Packet header shall contain a priority.

- *Scalability*

  - The RTOS shall support at least 64K tasks per node.
  - The RTOS shall support at least 64K nodes per system.
  - The maximum number of nodes shall be compile time defined.
  - The maximum of OpenComRTOS entities shall be compile time defined.
  - Shall support single processor targets with the same kernel as multiprocessor targets.
  - Packet management shall be done at system level.
  - The nodeID shall be part of the EntityID and shall be transparent for the application.

- *API conformant to RTOS*

  - The API shall support:

    · Task management
    · Events
    · Semaphores
    · Fifos
    · Resource locking
    · Memory management

  - The kernel implementation shall provide:

    · Priority based pre-emptive scheduling.
    · Priority inheritance support for resource entities.
    · Priority inheritance for tasks requesting a shared resource.
    · Time-out on service completion.

- *Extensibility*

  - The services shall be implemented independently from the kernel itself.
  - The programming environment shall support extensions without the need to rebuild the kernel.

- *Distributed operation*

  - The application code shall be independent of the nodes topology and entities mapping.

- The target system can be heterogeneous, linking processing nodes with different word lengths, using different compilers and connected using application specific communication systems.
- The target system can include third party host-OS.
- On such a node, the OpenComRTOS shall run on top of the host-OS.
- Routing shall be a system level functionality.
- Host services on third party host-OS shall be transparently accessible from any task on any node.

- *Efficiency*

  - The code size shall be less than 5 KB for single processor and less than 10 KB for a multi-processor implementations.
  - The system shall use a minimum amount of memory for packets and stack space.
  - The compiled code shall not link in unused services.
  - The kernel and drivers shall be implemented as application level tasks.
  - Compile time switches shall allow the generation of the smallest possible kernel for application development.

- *Safety*

  - No dynamic memory allocation shall be used in the kernel.
  - All datastructures are statically allocated at compile time.
  - OpenComRTOS shall use a separate stack space for handling interrupts.
  - When the packet pool is exhausted, the system shall not crash but start idling.
  - Drivers shall execute as application tasks.
  - The kernel shall have assertions enabled in the development version.

- *Runtime profiling*

  - An event trace shall be integrated.
  - The kernel shall allow read access of its state variables and datastructures.

- *Formally modelled and verified*

  - All kernel services shall be modelled and verified using a formal model checker.
  - All models shall verify with no reachable illegal states.
  - Model checking shall be used to verify the architectural design.
  - Model verification shall be done on the basis of the approved source code.

- *Automation*

  - A graphical environment shall be used to define the application.
  - A project shall be described in text file for manual or automated processing.
  - A meta-modelling concept shall be used to allow redefinition of entities and interactions.
  - It should be possible to run complete applications be recompilation in simulation mode on a host-OS.

- All code and datastructures shall be statically generated, unless application specific.
- The executing code shall be generated automatically and independently of the target system.

2.7   Systems and Application Grammar of OpenComRTOS

2.7.1   Base Principles and Definitions

An OpenComRTOS application can be constructed from various kernel entities and interactions between them, commonly called kernel services. Kernel entities and kernel services differ from one RTOS to another, but remain invariant at the meta-level of the interacting entities paradigm as instances of the RTOS System and application meta-model. They can be described in a "grammar", a textual representation comparable with an entity-relationship diagram. Such a grammar defines how correct programs can be constructed (at least at the syntactical level). The term system grammar is used as in languages whereby the grammar defines the rules that must be obeyed to create well-formed sentences. In this context, the sentence is a well formed real-time application or system and hence the system grammar is like a meta-model of the domain.

The meta-model relies on a typification of the Kernel entities and services. The base types of the interacting entities paradigm are as follows:

```
[Kernel Entity IS Entity.]
[Kernel Service IS Interaction.]
```

Next we can define the attributes:

```
[Kernel Entity HAS Attributes (2-N)     // e.g. name, node, type,
                                            parameters]
[Kernel Entity HAS Functions (0-N       // e.g. function]
[Kernel Interaction HAS Attributes (4) // name, subject, object,
                                            direction]
```

The purpose of the systems grammar is to define the OpenComRTOS system and application grammar as the minimal set of concepts needed for application development. This means that the internal mechanisms of how OpenComRTOS functions will be hidden from the application developer. For example, routing and scheduling are considered as internal OpenComRTOS kernel mechanisms and hence the entities and interactions that are involved in providing these services are not visible for an application developer.

There are two possible ways of modelling interactions:

1. If we define the Kernel Task as a separate entity we can consider kernel services as direct interactions between application Tasks and the Kernel Task.

2. If we hide the kernel entity i.e. consider it as an OpenComRTOS internal mechanism then all kernel services become interactions between application tasks by means of intermediate entities (e.g. hubs).

In the OpenComRTOS Open Visual Environment (OpenVE) the Kernel entity is hidden and the visual programming is done immediately at the application level. This is more natural as the underlying interactions between kernel entities and drivers can be complex whereas at the application level, the user is only interested in the end result.

In general, the concrete implementation of kernel services makes uses of packets exchanges for services calls or data transfer. Most of the time these packets also remain invisible.

In reality, we can differentiate five different types of OpenComRTOS Tasks: the Kernel Task, the Idle Task, RX Driver Tasks, TX Driver Tasks and Application Task. The application developer only needs to be concerned with application Tasks. This means that for an application development the Kernel Task is used as an indirect source of services for the application tasks. RX and TX Driver Tasks are defined at the level of the nodes topology and are never directly used by the OpenComRTOS application developer.

An OpenComRTOS application can be constructed from following base entities: Task, Port, Event, Semaphore, Resource, FIFO, Packet Pool and Memory Pool, each with associated services. Application development is a definition of interactions between Tasks through intermediate synchronisation entities, derived from a generic Hub (see further). The Task management services are considered as interactions between application Tasks (i.e. the object and the subject of interaction are of the same type).

2.7.2 A Note on Typing Conventions

In the rest of the book, we will use upper case words like Tasks, Hubs to indicate that we are talking about OpenComRTOS specific Entities. If the term is used in the general sense of the word, small case will be used.

2.7.3 OpenComRTOS System and Application Grammar

```
OpenComRTOS IS_DEFINED_BY
          SystemConfiguration (1) AND
          ApplicationConfiguration (1)

SystemConfiguration IS_DEFINED_BY
               SystemTasks (4) AND
               Topology (1)
```

```
SystemTask CAN_BE                    // Type of system task
          KernelTask OR              // Kernel itself
          IdleTask OR                // Background task
          RxTask OR                  // Link driver receiver task
          TxTask                     // Link driver transmitter task

SystemTask HAS_ATTRIBUTES
          EntryPoint (1) AND         // Function name
          Priority (0-1) AND         // Integer 1-255
          Arguments (0-N) AND        // Set of arguments or NULL pointer
          Status (0-1) AND           // Task status at creation
          StackSize (0-1)            // Size of static allocated stack

Topology IS_DEFINED_BY Nodes (1-N) AND
          Links (0-N)                // Real or virtual connection between
                                     // nodes

Node HAS_ATTRIBUTES Name (1) AND     // The Node Id
     TraceBufferSize (1) AND         // Size of trace buffer
     KernelPacketPoolSize (1) AND    // Default value 2
     RxPacketPoolSize (1) AND        // Default value 2
     Host (1)                        // The Name of the hosting Node

Link HAS_ATTRIBUTES Source (1) AND // Originating end-point
     Target (1) AND                  // Terminating end-point
     Type (1)                        // Bi- or unidirectional

LinkType CAN_BE Unidirectional OR
         Bidirectional

SystemConfiguration HAS_ATTRIBUTES      // XML node parameters
                    DataSize (0-1) AND  // Packet data size
                    NodeIdSize (0-1)    // Length of Node identifier

ApplicationConfiguration IS_DEFINED_BY ApplicationTasks (0-N) AND
                         Ports (0-N) AND        // Port hub
                         Event (0-N) AND        // Event hub
                         Semaphore (0-N) AND    // Semaphore hub
                         Resource (0-N) AND     // Resource hub
                         FIFO (0-N) AND         // FIFO hub
                         MemoryPool (0-N) AND   // Memory Pool hub
                         PacketPool (0-N) AND   // Packet Pool hub
                         Interactions (0-N)     // Hub service

ApplicationTask HAS_ATTRIBUTES Name (0-1) AND // Logical Name
                EntryPoint(1) AND              // Function name
                Priority (1) AND               // Integer 2-254
                Arguments (0-N) AND            // Set of arguments
                Status (1) AND                 // Task status
                Node (1) AND                   // Name host node
                StackSize (1)                  // Size allocated stack
```

```
TaskStatus CAN_BE L0_INACTIVE OR
           L0_STARTED                            // Default

Port HAS_ATTRIBUTES Name (1) AND
     Node (1)

Event HAS_ATTRIBUTES Name (1) AND
      Node (1)

Semaphore HAS_ATTRIBUTES Name (1) AND
          Node (1)

FIFO HAS_ATTRIBUTES Name (1) AND
     Node (1) AND
     Size (1)                                    // Default value 1

Resource HAS_ATTRIBUTES Name (1) AND
         Node (1) AND

MemoryPool HAS_ATTRIBUTES Name (1) AND
           Node (1) AND
           NumOfBlocks (1) AND                   // Default value 1
           SizeOfBlocks (1)                      // Default value 1024

PacketPool HAS_ATTRIBUTES Name (1) AND
           Node (1) AND
           Size (1)                              // Default value 1

Interaction HAS_ATTRIBUTES Name (1) AND
            Subject (1) AND
            Object (1) AND
            Service (1)                          // either a Put or
                                                 // Get interaction

InteractionService CAN_BE Put OR Get
```

This system grammar can be considered as a formalised specification for the OpenComRTOS implementation. Its elements are found back in the meta-models used by the application development environment and the code generators. For the list of specified interaction services, omitted from the systems grammar for reason of compactness, we give the corresponding extract from the meta-models in XML format.

```
<! Acromyms:>
<! Int  = Interaction >
<! Subj = Subject>
<! Obj  = Object>
<! Serv = Service>

<!-- Task management Services -->
<Int ="L1_StartTask_W" Subj="Task" Obj="Task" Serv="Put"/>
<Int ="L1_StopTask_W" Subj="Task" Obj="Task" Serv="Put"/>
```

```xml
<Int ="L1_SuspendTask_W" Subj="Task" Obj="Task" Serv="Put"/>
<Int ="L1_ResumeTask_W" Subj="Task" Obj="Task" Serv="Put"/>
<Int ="L1_SleepTask_WT" Subj="Task" Obj="Task" Serv="Put"/>

<!-- Port based Services -->
<Int ="L1_PutPacketToPort_W" Subj="Task" Obj="Port" Serv="Put"/>
<Int ="L1_GetPacketFromPort_W" Subj="Task" Obj="Port" Serv="Get"/>
<Int ="L1_PutPacketToPort_NW" Subj="Task" Obj="Port" Serv="Put"/>
<Int ="L1_GetPacketFromPort_NW" Subj="Task" Obj="Port" Serv="Get"/>
<Int ="L1_PutPacketToPort_WT" Subj="Task" Obj="Port" Serv="Put"/>
<Int ="L1_GetPacketFromPort_WT" Subj="Task" Obj="Port" Serv="Get"/>

<!-- Event based Services -->
<Int ="L1_RaiseEvent_W" Subj="Task" Obj="Event" Serv="Put"/>
<Int ="L1_TestEvent_W" Subj="Task" Obj="Event" Serv="Get"/>
<Int ="L1_RaiseEvent_NW" Subj="Task" Obj="Event" Serv="Put"/>
<Int ="L1_TestEvent_NW" Subj="Task" Obj="Event" Serv="Get"/>
<Int ="L1_RaiseEvent_WT" Subj="Task" Obj="Event" Serv="Put"/>
<Int ="L1_TestEvent_WT" Subj="Task" Obj="Event" Serv="Get"/>

<!-- Semaphore based Services -->
<Int ="L1_SignalSemaphore_W" Subj="Task" Obj="Semaphore" Serv="Put"/>
<Int ="L1_TestSemaphore_W" Subj="Task" Obj="Semaphore" Serv="Get"/>
<Int ="L1_SignalSemaphore_NW" Subj="Task" Obj="Semaphore"
     Serv="Put"/>
<Int ="L1_TestSemaphore_NW" Subj="Task" Obj="Semaphore" Serv="Get"/>
<Int ="L1_SignalSemaphore_WT" Subj="Task" Obj="Semaphore"
     Serv="Put"/>
<Int ="L1_TestSemaphore_WT" Subj="Task" Obj="Semaphore" Serv="Get"/>

<!-- Resource related Services -->
<Int ="L1_LockResource_W" Subj="Task" Obj="Resource" Serv="Get"/>
<Int ="L1_UnlockResource_W" Subj="Task" Obj="Resource" Serv="Put"/>
<Int ="L1_LockResource_NW" Subj="Task" Obj="Resource" Serv="Get"/>
<Int ="L1_UnlockResource_NW" Subj="Task" Obj="Resource" Serv="Put"/>
<Int ="L1_LockResource_WT" Subj="Task" Obj="Resource" Serv="Get"/>
<Int ="L1_UnlockResource_WT" Subj="Task" Obj="Resource" Serv="Put"/>

<!-- FIFO Queue related services -->
<Int ="L1_EnqueueFifo_W" Subj="Task" Obj="FIFO" Serv="Put"/>
<Int ="L1_DequeueFifo_W" Subj="Task" Obj="FIFO" Serv="Get"/>
<Int ="L1_EnqueueFifo_NW" Subj="Task" Obj="FIFO" Serv="Put"/>
<Int ="L1_DequeueFifo_NW" Subj="Task" Obj="FIFO" Serv="Get"/>
<Int ="L1_EnqueueFifo_WT" Subj="Task" Obj="FIFO" Serv="Put"/>
<Int ="L1_DequeueFifo_WT" Subj="Task" Obj="FIFO" Serv="Get"/>

<!-- Packet Pool services -->
<Int ="L1_AllocatePacket" Subj="Task" Obj="PacketPool" Serv="Get"/>
<Int ="L1_AllocatePacket_W" Subj="Task" Obj="PacketPool" Serv="Get"/>
<Int ="L1_DeallocatePacket_W" Subj="Task" Obj="PacketPool"
     Serv="Put"/>
```

```
<Int ="L1_AllocatePacket_NW" Subj="Task" Obj="PacketPool"
     Serv="Get"/>
<Int ="L1_AllocatePacket_WT" Subj="Task" Obj="PacketPool"
     Serv="Get"/>

<!-- Memory Pool related Services -->
<Int ="L1_AllocateMemoryBlock_W" Subj="Task" Obj="MemoryPool"
     Serv="Get"/>
<Int ="L1_DeallocateMemoryBlock_W" Subj="Task" Obj="MemoryPool"
     Serv="Put"/>
<Int ="L1_AllocateMemoryBlock_NW" Subj="Task" Obj="MemoryPool"
     Serv="Get"/>
<Int ="L1_AllocateMemoryBlock_WT" Subj="Task" Obj="MemoryPool"
     Serv="Get"/>
```

Part II
Formal Modeling Fundamentals

Chapter 3
The Choice of TLA$^+$/TLC: Comparing Formal Methods

This book provides a thorough description of OpenComRTOS and of the formal models built for its implementation. Such models were written in TLA$^+$ and verified with the TLC model checker. The choice of TLA$^+$/TLC was one of the fundamental initial decisions of the project. In effect, over the last years, several languages, techniques, and tools have been developed and made available by the formal methods community. Typically, different formalisms are best suited for different domains of application, and their comparison is not straightforward. It can be a significantly subjective exercise. This chapter addresses this issue, describing the selection process followed in this project.

We identified a set of criteria for evaluation and applied the selected formalisms to a common case study: a non-blocking algorithm for managing linked-lists. There was no particular reason to select this algorithm as a test case, except that it is of potential interest in the context of RTOS kernels that often manipulate linked lists. After a survey and pre-selection phase, the languages and corresponding tools analysed were TLA$^+$/TLC and Promela/SPIN. The chapter first describes the pre-selection stage, then describes the algorithm in detail, the selected formalisms, and, finally, the comparison criteria and evaluation results.

3.1 Formal Methods Survey and Pre-Selection

In classical engineering disciplines (electrical, mechanical, etc.) tools like Maple and MatLab have a wide scope of utilization and are widely used. This is not yet the case for tools supporting the formal development of software systems. Instead, there is a rather large set of languages, methods, and tools with different and more or less restricted domains of application.

Formal specification languages can be classified with respect to their prime application domain, i.e., either specialized towards sequential programs, or focusing on concurrent and distributed systems. Well established and representative examples

E. Verhulst et al., *Formal Development of a Network-Centric RTOS: Software Engineering for Reliable Embedded Systems*, DOI 10.1007/978-1-4419-9736-4_3,
© Springer Science+Business Media, LLC 2011

of the first are B, VDM, and Z; whereas CCS, CSP, TLA$^+$, Unity and Petri Nets are mainly used to specify and reason about concurrent systems.

The correctness of a system is assessed by confronting the specification of the system/algorithm with the statement of the properties desired and checking whether one implies the other. This can be done by writing proofs by hand or using more or less automated tools. Two main groups of tools can be considered: proof systems and model checkers.

Proof systems generally require some user interaction. They can check whether a given proof is valid, according to the rules of deduction of the logic concerned, or, given a theorem and its premises, try to discover a proof such that the theorem follows from the premises. An hybrid approach can also be utilized: the user proposes large steps and the systems fills in the gaps and takes care of details. Examples of proof systems are Coq, HOL, Isabelle, Nuprl, PVS, and B-related tools.

Model checkers are specially suited for analyzing (descriptions of) concurrent systems. They can differ in many aspects, namely, whether they use branching or linear temporal logic, symbolic or explicit state verification, breadth-first search or depth-first search, real-time or timeless verification, etc.

Examples of symbolic model checkers are SMV and NuSMV. Murphi, SPIN and TLC are explicit-state model-checkers. Kronos and UPPAAL focus on timed systems. Other relevant model checkers are μCRL and FDR. Model checkers more frequently take as input specifically developed languages. Relevant exceptions are FDR, for checking CSP specifications, and TLC (for TLA$^+$ specifications).

Because tool support is of great importance and since most of the problems that had to be analyzed are related to concurrency issues, model checking is a particularly suitable technique.

Depending on their design decisions, model checkers can be more appropriate to verify hardware or software systems. (Naturally, the choice should fall on a model checker primarily intended for software systems.) SMV, NuSMV, and Murphi are mainly intended for hardware descriptions verification, which lessens their suitability. Kronos and UPPAAL are used to verify real-time systems. However, they take as input rather low-level languages and time constraints are not well addressed. Moreover, logical correctness and timing requirements are orthogonal issues. SPIN and TLC are model checkers of widespread utilization. Their input languages, Promela and TLA$^+$, provide large support for specifying concurrent systems. The two seemed a good fit for the purpose of the project. At least at the time of starting this project, because since then research has continued which could have resulted in a different decision today.

3.2   Case Study

The case study selected was Harris' "pragmatic implementation of non-blocking linked-lists" (Harris 2001). In this section, we present a textual description of the algorithm, complementary to the one in the cited reference, together with an analysis of the potential problems of the subject.

3.2.1   Introduction

Linked-lists are a basic structure used in program design. A single linked list can be simply defined as a chain of elements which contain a pointer to the next element. Each element is commonly referred to as a *node*. Each node contains two fields: a *key* field, identifying the node, and a *next* field, containing a reference (pointer) to the next node in the list. The two basic operations that a process can do are (a) inserting and (b) deleting a node to/from the list. Both operations comprise locating the place where to insert/delete the node to/from the list and then *physically* perform the operation. A node is inserted by, after identifying its predecessor and successor nodes, making it point to the successor and changing the predecessor's reference to point to the new node. Deleting a node can simply be achieved by making its predecessor point to its successor, eliminating the references to itself.

Several problems with concurrent access to the list can arise when, e.g. (a) different processes try to insert nodes with the same identifier, (b) different processes try to delete nodes with the same identifier, (c) different processes try to insert nodes with different identifiers but the same predecessor and successor, and (d) a process tries to insert a node whose identified successor/predecessor is being deleted by another process. In more detail, for the last scenario, take for example the case where two concurrent processes are, respectively, trying to insert and delete a node. These nodes are such that the node to be inserted has the node to be deleted as predecessor. If the inserting process terminates first but still late enough that the deleting process *didn't see it* when it was identifying its successor and predecessor nodes, when the node is deleted the new one that had just been inserted will be lost, causing a violation of the intended result.

Traditionally, these and similar problems are solved by granting exclusive access to the shared resources (the linked list in this case), which prevents all other processes trying to access the same resource from making progress. By contrast, an implementation is non-blocking (or lock-free) if some process must complete an operation after the system as a whole takes a finite number of steps, which means that some process will always make progress despite arbitrary halting failures or delays by other processes (Herlihy 1993). The algorithm presented by Harris (2001) is able to deal with linked-list in a non-blocking fashion.

3.2.2   The Algorithm

Harris algorithm makes use of two fundamental features. The first is to use an atomic compare-and-swap (CAS) primitive. CAS (addr, old, new) atomically compares the contents of a location *addr* with an expected value *old* and, if they match, writes the value *new* to that location. CAS returns a boolean value indicating whether the substitution took place or not.

The second is to separate the deletion of a node in two phases. The first to *mark*[1] the node and the second one to remove it from the list. The node is considered *logically deleted* after the first stage and *physically deleted* after the second. A marked node is still part of the chain of the linked list, but its marking is used to signal concurrent processes to not introduce new nodes immediately after those that are logically deleted.

With the CAS operation the problems identified above as (a) and (c) can be solved straightforwardly. The *physical* insertion of a node is only achieved when the reference of its predecessor is changed. This is done atomically with CAS. If, at the moment when the instruction is executed, some other process made a *significant* change to the list, CAS will fail, and the process restarts by relocating again its predecessor and successor nodes. The separation of the deletion in two phases allows the overcoming of the problems identified as (b) and (d).

To simplify the management of the list, a typical procedure is used: two special root nodes, *"Head"* and *"Tail"* are added. *Head* is always kept as the first element of the list and *Tail* as the last one. The insertion and deletion are then defined as follows.

Insertion Procedure:

(Ii) The node to be inserted *current* is generated;

(IIi) The predecessor *left* and successor *right* nodes are identified (all the nodes have a key and they are stored in the list in ascendant order). The addresses of these nodes are stored in two variables. The *left* node is the unmarked node that has the biggest key strictly smaller than the *current* node key. The *right* node is the unmarked node that has the smallest key greater or equal than the *current* node key.

(IIIi) The uniqueness of the node to insert is checked: if the key of the *right* node is equal to the one of the *current* node, the process is aborted; otherwise it continuous its execution.

(IVi) The *current* node is made to point to the *right* node (by a simple assignment instruction)

(Vi) The *physical insertion* of the node, utilising the CAS (adr, old, new) operation is attempted: the node that is being presently pointed at that instant by the left node *[addr]* is compared to the one previously identified *[old]*, whose value has been stored in a variable and:

  (i) If they match, the *left* node is made to point to the *current* node, concluding the inserting process.

  (ii) If they do not match (which means that some change in the list has been made in the mean time), the process goes back to (IIi).

---

[1] The actual implementation of the marking can be done using an unused bit of the field *next* of the node.

3.2.2.1   Deletion Procedure

(Id)   An identifier key *value* is selected.

(IId) The list is searched in order to identify the *left* and *right* nodes. The rule is as mentioned just above the *left* is the unmarked node that has the biggest key strictly smaller than *value*; the *right* is the unmarked node that has the smallest key greater or equal than *value*.

(IIId) The identity of the right node is checked: if its key is equal to *value* the process continues to execute and the right node is the node to be deleted; otherwise it aborts.[2]

(IVd) A new *neighborhood node* is identified: the immediate successor of the right node is stored in a variable *right_next*.

(Vd)   The *marking* of the (right) node is attempted (atomically, with a CAS operation): if the right node next field *[addr]* is still pointing to the node previously identified *right_next [old]*, then its value is rewritten as marked *[new]*. If not the process goes back to (IId). This operation is guarded by the pre-condition that *right_next* is not a marked node; otherwise the process goes directly back to (IId).

(VId) The *physical deletion* of the (right) node is attempted (again atomically, using the CAS operation): if the left node next field *[addr]* is still pointing to the right node *[old]*, then it is made to point to the node identified as its immediate successor *right_next [new]*. Once again if this fails the process goes back to (IId).

3.2.3   Remarks

In the original algorithm, the procedure for identifying the left and right nodes – (IIi) and (IId) – is more elaborate than the description given above. It is executed using a *search* routine that incorporates a CAS operation to remove marked nodes between the left and right nodes, before returning. The simplification introduced here allows a simplification of the models to develop, while keeping faithful to the core behaviour of the original algorithm.

3.2.4   Drawbacks

In Harris algorithm, a node may still be accessed after being deleted from the list. In fact, in some scenarios, this may be necessary. However, it prevents the

---

[2]This mail happen for two reasons: (a) another concurrent process could have deleted the node in the mean time or (b) there was some error in the assignment of the node to delete and it simply does not belong to the linked list.

algorithm from using simple and efficient lock-free memory management methods. Let's illustrate the problem with an example.

With memory release:

(1) Process 1 is assigned to delete a node with key *k (Id)*, that belongs to the linked list.

(2) Process 1 identifies the neighborhood of the node *n (IId, IIId, IVd)*.

(3) Process 2 is assigned to delete a node with the same key *k*.

(4) Process 2 identifies the same neighborhood of the node *n (IId, IIId, IVd)*.

(5) Process 1 marks the node *n (Vd)*.

(6) Process 1 excises the node *n (VId)*.

(7) Process 2 tries to mark the node *n (Vd)* but, due to Process 1 releasing the memory of node *n(Vd)* its memory location is now empty, causing an error.[3]

Without (or "with sufficiently delayed") memory release:

(1)–(6) Same as before.

(7) Process 2 tries to mark node *n (Vd)*, but the comparison of the CAS operation fails[4], and the process is sent back to (IId).

(8) Process 2 identifies the new left and right nodes *(IId)*.

(9) Two different things may happen now: (a) the key of the new right node is bigger than *k* and the process ends its execution *(IIId)*, or (b) if, in the mean time, another process has inserted back a node with that same key *k*, process 2 will detect it and continue the normal deletion procedure.

3.2.5   Related Work

Due to the natural advantages of lock-free implementations over the traditional use of locks, extensive work has been done on this subject. Atomic primitives like CAS play a central role in this kind of algorithms and modern processors support them at hardware level.[5] Another primitive commonly used is the *Load-Link/Store-Conditional/Validate* (LL/SC/VL) (Jayanti and Petrovic 2004; Michael 2004a; Moir 1997).

---

[3]"Empty" is not an exact term. Once the memory location has been freed it can be in a multiplicity of different states that may lead to the referred error. The actual behavior of a process in this scenario depends on the details of the implementation.

[4]Recall that the comparison is made between the value of the node's next field previously recorded and the present one. Even if the composition of the list has not changed in the mean time, the values will not match because the present reference is marked while the value previously recorded is not. This is due to decision of marking a node by a change in an unless unused bit of the next filed of the node. In any case, if other implementation scheme would not cause this comparison to fail the process would continue to (VId) and would then forcedly be sent back to (IId), continuing as announced in (8).

[5]For example, the CAS primitive is supported as the "CASA" and "CASXA" operations in Sparc V9 (SPARC International 1994), and as "lock cmpxchg" in Intel Itanium (Intel Corporation 2002).

In (Michael 2002a), it is pointed out that the memory management[6] issues in Harris' algorithm are a serious drawback. The same author presents, in (Michael 2002b), what he calls "the reclamation memory" method as a solution.

A problem that affects many lock-free algorithms is the so-called "ABA problem" (IBM 1983). The atomic primitives LL/SC/VL seem to be a more convenient way of developing ABA problem free algorithms (Michael 2004b). An algorithm immune to this problem using LL/SC/VL is presented in (Michael 2004a). A solution with CAS is given in (Gao and Hesselink 2004).

3.3 TLA$^+$ and TLC

3.3.1 Overview

Formal model checking techniques often consist of a formal language and a model checker. TLA$^+$ is a specification language for describing and reasoning about asynchronous, nondeterministic, concurrent systems and TLC is an explicit-state model checker for specifications written in TLA$^+$. An introduction to the main concepts of TLA$^+$ will be given in Chap. 4. Here, we give an introduction to the details of checking TLA$^+$ specifications with TLC.

We take as a running example the model shown in Fig. 3.1. It is a very simple model that manages a list of elements (keys). The possible operations are to insert and to delete elements from the list. The list is structured as a sequence. Insertion is done by appending the new element to the bottom of the sequence. Deletion consists in simply removing the first element from the top of the list.

The specification starts with

┌──────────────────── MODULE *Intro* ────────────────────┐

that begins a module called *Intro*. As mentioned, TLA$^+$ specifications can be partitioned into multiples modules. We use only one in this example.[7]

Line 2 introduces EXTENDS, to incorporate the *Naturals* and *Sequences* modules. The *Naturals* module contains the definitions of the "usual operators on natural numbers", like '+' or '>'. The *Sequences* module will be described in Chap. 4.

Line 3 introduces the constant parameters *Keys* and *N*. *Keys* is a set used to define the elements that can be part of the list. *N* defines the maximum length of the list. The model contains only one variable, *list*. (Again, please note that TLA$^+$

---

[6]See Sect. 3.2.4.

[7]Together with the standard modules *Naturals* and *Extends*.

```
 1 ┌──────────────────────── MODULE Intro ────────────────────────┐
 2   EXTENDS Naturals, Sequences
 3   CONSTANT Keys, N   Keys and N are both defined in the config file
 4   VARIABLES list
 5   ASSUME (N ∈ Nat) ∧ (N > 0)
 6   TypeInvariant  ≜  list ∈ Seq(Keys)
 7 ├──────────────────────────────────────────────────────────────┤
 
 9   Init ≜ list = ⟨⟩      List initialized empty
 
12   Insert(key) ≜  ∧ Len(list) < N
13                  ∧ list' = Append(list, key)
 
16   Delete ≜  ∧ Len(list) > 0
17             ∧ list' = Tail(list)
 
 
21   Next  ≜  ∨ ∃k ∈ Keys : Insert(k)
22            ∨ Delete
 
24   Spec  ≜  Init ∧ □[Next]_list
25 ├──────────────────────────────────────────────────────────────┤
26   THEOREM Spec ⟹ □ TypeInvariant
27 └──────────────────────────────────────────────────────────────┘
```

Fig. 3.1 Example of a specification in TLA$^+$

is an untyped language: the variables are simply listed, not typed.) The clause ASSUME $(N \in Nat) \wedge (N > 0)$ asserts that we are assuming that N is a positive natural number.[8]

$$TypeInvariant \quad \triangleq \quad list \ \in Seq(Keys) \qquad \text{(Line 6)}$$
defines the correctness criteria of the model: at any state, the list has to be a valid sequence whose elements all belong to the set *Keys*.

The line

├──┤

is purely cosmetic, it can appear anywhere.

Once explained, the *header* of a TLA$^+$ module, a specification is better understood if we read the rest of it from the bottom to the top. At this stage, going through it should be straightforward. Line 26 expresses the correctness claim that we want TLC to verify. Moving up, *Spec* is the specification of our program. It tell us that our system is a state machine, whose initial state is *Init* and that can evolve through the steps defined in *Next*, which is defined as the disjunction of the actions *Insert* and *Delete*. The list is initialized empty.

---

[8]This has no effect in the definitions made in the module; it can, however, be taken as an hypothesis in the construction of proofs.

For a nice and easy to read presentation, all comments in a TLA$^+$ module are shaded. A module terminates with the symbol

When verifying a model with TLC we need to explicitly define the system constants. This is due to the fact that TLC (like SPIN) only handles systems with a finite number of states. The definition is done in a separate configuration file. Listing 3.1 below illustrates a possible configuration file for this example.

```
1   (*******************************************************************)
2   (*                      CONFIG  Intro.cfg                       *)
3   (*******************************************************************)
4
5   SPECIFICATION Spec
6   (* This statement tells TLC that Spec is the specification to be  *)
7   (* checked.*)
8
9   INVARIANT TypeInvariant
10  (* This statement tells TLC to check that TypeInvariant is an     *)
11  (* invariant of the specification.                                *)
12
13  (*******************************************************************)
14  (* TLC requires that every declared constant in the specification *)
15  (*be assigned a value by a CONSTANT statement in the configuration*)
16  (*file.                                                           *)
17  (*******************************************************************)
18  CONSTANT
19  N = 3
20  Keys = {10, 20, 30}
21  (*********************** End of config file  ********************)
```

Listing 3.1 Configuration file

TLC can be run in *model checking* or *simulation* mode. In model checking mode, it tries to find all the reachable states. It is the default mode. In simulation mode, it randomly generates behaviors up to a specified maximum length.

3.3.2 Model Developed

In this section, we present the developed TLA$^+$ model of Harris algorithm. The design decisions are explained and some extracts of the model are shown. The full model is given in Appendix C.1 on Page 199.

The central aspect of writing a (TLA$^+$) specification is choosing the level of abstraction. This implies choosing the variables that represent the system and choosing the granularity of the steps that allow the system to evolve. The granularity chosen is very simple to summarize: it basically follows the textual

description made in Sect. 3.2.2, where each of the phases identified as Ii .. Vi and
Id .. VId corresponds to an atomic action in the TLA$^+$ model.[9]

Two variables were defined: *mem* and *proc*. Variable *mem* represents
the memory model of the system: a series of memory addresses where nodes
can be recorded. The list of addresses is enumerated in the constant *Adr* and the
nodes are represented as a record of three fields: *key*, *next* and *mark*. The set
of all possible keys is defined in *Keys*. The *next* field (when not empty) should
contain an address – to represent the pointer of a node to its successor. *mark* is a
boolean variable – a node can only be marked or not. *mem* is defined as a function
of addresses to nodes.

The variable *proc* encapsulates the information to deal with the behaviour of
the system. This information is, on its turn, encapsulated in four records: *ninfo*,
procIns, *procDel* and *choice*. The record *ninfo* registers the information that
effectively needs to kept in between intermediate steps, i.e. the key and next values
of the node to insert/delete and its neighborhood.

```
1   (*****************************************************************)
2   (*                       CONFIG  HarrisR.cfg                    *)
3   (*****************************************************************)
4
5   SPECIFICATION Spec
6   (* This statement tells TLC that Spec is the specification to be *)
7   (* checked.                                                      *)
8
9   INVARIANTS TypeInvariant Coherence
10  (* This statement tells TLC to check that TypeInvariant and      *)
11  (* Coherence are invariants of the specification.                *)
12
13  (*****************************************************************)
14  (* TLC requires that every declared constant in the specification *)
15  (*be assigned a value by a CONSTANT statement in the configuration*)
16  (*file.                                                          *)
17  (*****************************************************************)
18  CONSTANT
19     Keys    = {10,20,30,40,50}
20     Adr     = {1201,1202,1203,1204,1205,1206,1207,1208}
21     Process = {p1,p2,p3,p4}
22     HEAD    = 1
23     TAIL    = 100
24  (******************** End of config file  ********************)
```

Listing 3.2 Configuration file of Harris Algorithm TLA$^+$Model

The records *procIns*, *procDel* and *choice* represent "auxiliary variables" of
the model, i.e. they are used to control aspects that are not directly connected to
the algorithm, but to the TLA$^+$ model itself. Because the insertion and deletion
processes were both modeled in several different actions, an extra mechanism
is needed to ensure the correct sequence of execution. This was done using to

---

[9]Strictly speaking, there is an exception: IId and IIId were grouped into a single action.

a variable that registers in what intermediate step the execution is. *procIns* is the variable for the insertion process and *procDel* for the deletion. The variable *choice* identifies the commitment of a process (after being committed a process is only set free when all intermediate steps have terminated).

Because we want to have a system with multiple processes running, the variable *proc* is made a function of processes that refers to the four records of this process (*ninfo*, *procIns*, *procDel* and *choice*). The set of existing processes is defined in the constant *Process*. Figure 3.2 shows the header of the model. Listing 3.2 shows a possible instance of the configuration file.[10]

Once we have understood the variable abstraction, we should move to the bottom of the module (Fig. 3.3). Its reading is straightforward. The theorem of the model is that our specification implies the invariants *TypeInvariant* and *Coherence*[11]. The invariant *Coherence* expresses the order relation that must be kept in between the nodes. It states that all the nodes registered in *mem* and that point to another node, have a *key* that is strictly smaller than the key of the node that they point to (Fig. 3.4).

The specification *Spec* of the system is defined by the initial conditions expressed in *Init* and the next-state relation *Next*. *Next* is a disjunction of the actions *SetInitNodes*, *Insert* and *Delete*. *SetInitNodes* is the "auxiliary action" where the special root nodes *Head* and *Tail* are inserted into the list. *Insert* and *Delete* both take as parameters a *process* (responsible for the execution) and a *key* (identifying the node to insert/delete). Both actions are defined as a disjunction of five finer grained steps. Table 3.1 shows the correspondence in between them and the textual description made in Sect. 3.2.2.

As a representative example of all the other actions, we'll cover the definitions of *CreateI*, *LocateD* and *CasD2*. *CreateI*(*p*, *key*) in Fig. 3.5, which takes as parameters a process *p*, to execute the action, and the *key* of the node that we want to insert. It is guarded by the conditions that the *setup* (insertion of *Head* and *Tail*) was already done (*setup* = 1) and that the process *p* is ready to start and not yet committed to any other action: *proc*[*p*].*procIns* = "ready!" ∧ *proc*[*p*].*choice* = "undecided". An extra "auxiliary" guard is added, stating that there has to be free memory: the number of free memory locations is greater than the number of processes executions insertions. The function used here, *Cardinality*, is defined in the standard module *FiniteSets*.

The actualization of a variable in TLA$^+$ may be done in two different forms: (1) explicitly stating the values of all its fields, or (2) by saying that is the same as before, except for the fields listed after EXCEPT [12]. The actualizations in *CreateI* update *choice*, *procIns* and register the input parameter *key* as the key of the current node to insert.

---

[10]Note the use of "possible instance". In fact, the configuration file may have many instances by changing the definitions of the constants in it. The TLA$^+$ model remains, however, unchanged.

[11]More rigorously, it asserts that the formula follows logically from the definitions in this module, the definitions in the extended modules *Naturals*, *Sequences*, and *FiniteSets*, and the rules of TLA$^+$.

[12]In our case, this feature is really useful since variable *proc* encapsulates many fields.

```
 1 ┌─────────────────── MODULE HarrisR ───────────────────┐
      This model represents the algorithm presented by Harris for the implementation of non-blocking
      linked-lists.

      In this model the nodes inserted to the list are stored in the global variable "mem". The actions of
      deleting and inserting nodes to the list are divided, for a finer-grained solution in several interme-
      diate steps.

      For 'artificial' (to be seen in some comments bellow) we mean all the aspects that are not directly
      connected to the algorithm, but that represent adaptations of it to be possible/more easily modeled.
10 EXTENDS Naturals, Sequences, FiniteSets
11 CONSTANT Adr,              the set of addresses
12            Keys,            the set of keys
13            Process,         the set of processes
14            HEAD, TAIL    values of the keys of 'Head' and 'Tail'

16 VARIABLES mem,     'state of the memory', with all the nodes inserted in the Linked List
17            proc,     auxiliary variable – information for the intermediate stages
18            setup     'artificial' variable for the initial insertion of 'Head' and 'Tail'

20 ASSUME    HEAD has to be smaller than any element of the set Keys, and TAIL has to be bigger
21            ∀k ∈ Keys : HEAD < k  ∧  k < TAIL
22 ├──────────────────────────────────────────────────────┤

25 TypeInvariant ≜    mem stores all the nodes inserted into the list. It's a function that assigns
26                     nodes to addresses. Each node as a key, a next field pointing to the adress of
27                     the next node, and can be marked or not.
28                     ∧ mem ∈ [Adr → [key : Keys ∪ {0, 1, 100},
29                                       next  : Adr ∪ {0},
30                                       mark : {0, 1}]]

32                     "setup" is 0 before the insertion of Head and Tail into the list, and 1 after that.
33                     ∧ setup ∈ {0, 1}

35                     "proc" is a function of each process
36                     ∧ proc ∈ [
37                        Process → [  ninfo is a record that keeps track of the information regarding
38                        the current node to be inserted/deleted, like its key and position in the list.
39                                      ninfo : [CNkey : Keys ∪ {0}, CNnext : Adr ∪ {0},
40                                                AdrLeft : Adr ∪ {0}, AdrRight : Adr ∪ {0},
41                                                RigNext : Adr ∪ {0}],

43                                      procIns and procDel state the finer-grained steps of the insert
44                        and delete actions
45                                      procIns : {"readyI", "createdI", "locatedI", "unique",
46                                                  "swapedI1"},

48                                      procDel : {"readyD", "identifiedD", "locatedD",
49                                                  "assignedD", "swapedD1"},

51                                      A process can only start inserting/deleting a new node if it's
52                        still free. Once it's committed to a certain action it should carry it till the
53                        end. "choice" is used to represent that. A simple 2 bit variable, e.g. {"free",
54                        "committed"}, would be a more elegant solution. The distinction between 'insert'
55                        and 'delete' is necessary for 'artificial reasons' – see step "CreateI(p, key)".
56                                      choice : {"undecided", "toinsert", "todelete"}]]
```

Fig. 3.2 *Header* of the TLA⁺ model

324 |———|

329   $Insert(i, k) \overset{\Delta}{=} CreateI(i, k) \vee LocateI(i) \vee VerUniq(i) \vee CasI1(i) \vee CasI2(i)$

331   $Delete(i, k) \overset{\Delta}{=} Identify(i, k) \vee LocateD(i) \vee AssignD(i) \vee CasD1(i) \vee CasD2(i)$

334   $Next \overset{\Delta}{=} \vee SetInitNodes$

336                $\vee \exists i \in Process : \exists k \in Keys : Insert(i, k) \vee Delete(i, k)$

339   $Spec \overset{\Delta}{=} Init \wedge \square[Next]_{\langle mem, proc, setup \rangle}$

340 |———|

341   THEOREM $Spec \implies \square(TypeInvariant \wedge Coherence)$

342 |———|

Fig. 3.3 Bottom of the TLA$^+$ Specification of Harris' algorithm

59   $Coherence \overset{\Delta}{=}$ "The key of the node that one node points to has to smaller than its own key"

60   LET   set of all nodes pointing to another one

61                      $nodp \overset{\Delta}{=} \{j \in Adr : (mem[j].key \neq 0 \wedge mem[j].next \neq 0)\}$

62   IN   Claim:

63                      $\forall i \in nodp : mem[i].key < mem[mem[i].next].key$

Fig. 3.4 Definition of *Coherence*

Table 3.1 Correspondence of TLA$^+$ model with the textual description in Sect. 3.2.2

Insert		Delete	
TLA$^+$ model	Textual description	TLA$^+$ model	Textual description
CreateI	Ii	Identify	Id
LocateI	IIi	LocateD	IId & IIId
VerUniq	IIIi	AssignD	IVd
CasI1	IVi	CasD1	Vd
CasI2	Vi	CasD2	VId

109   $CreateI(p, key) \overset{\Delta}{=} \wedge setup = 1$

110                    $\wedge proc[p].procIns = $ "readyI"

111                    $\wedge proc[p].choice = $ "undecided"

112   Checks if there's still space in memory

113                    $\wedge (Cardinality(\{a \in Adr : mem[a].key = 0\}) - Cardinality($

114                        $\{i \in Process : proc[i].choice = $ "toinsert"$\})) > 0$

116                    $\wedge proc' = [proc$ EXCEPT $![p].ninfo.CNkey = key,$

117 $![p].procIns = $ "createdI",

118 $![p].choice = $ "toinsert"$]$

119                    $\wedge$ UNCHANGED $\langle mem, setup \rangle$

Fig. 3.5 Definition of action *CreateI*

237   $LocateD(p) \triangleq \wedge proc[p].procDel =$ "identifiedD"

239                    $\wedge$ LET $posr \triangleq \{j \in Adr : (mem[j].key \neq 0 \wedge mem[j].mark = 0 \wedge$
240 $mem[j].key \geq proc[p].ninfo.CNkey)\}$

242                         $posl \triangleq \{j \in Adr : (mem[j].key \neq 0 \wedge mem[j].mark = 0 \wedge$
243 $mem[j].key < proc[p].ninfo.CNkey)\}$

245                         $right \triangleq$ CHOOSE $a \in posr : \forall i \in posr : mem[a].key \leq mem[i].key$

247                         $left \triangleq$ CHOOSE $a \in posl : \forall i \in posl : mem[a].key \geq mem[i].key$

249   IN
250   Check the identity of the right node: if its key is equal to the current one the process
251   continuous to execute and the right node is the node to be deleted; otherwise it aborts.
252   IF $mem[right].key = proc[p].ninfo.CNkey$ THEN

254                              $\wedge proc' = [proc$ EXCEPT $![p].ninfo.AdrLeft = left,$
255 $![p].ninfo.AdrRight = right,$
256 $![p].procDel =$ "locatedD"]
257                         $\wedge$ UNCHANGED $\langle mem, setup \rangle$

259   ELSE   abort (key doesn't exist)
260                              $\wedge proc' = [proc$ EXCEPT $![p].ninfo.CNkey = 0,$
261 $![p].procDel =$ "readyD",
262 $![p].choice =$ "undecided"]
263                         $\wedge$ UNCHANGED $\langle mem, setup \rangle$

Fig. 3.6 Definition of action *LocateD*

Action $LocateD(p)$, in Fig. 3.6, identifies the left and right nodes and checks
the *key* of the right node: if it is equal to the search key, right node is assigned to
be deleted. Otherwise (the node that we want to delete is not in the linked-list) the
process is set free ($procDel =$ "readyD", *choice* = "undecided").

The definition of *LocateD* introduces the TLA$^+$ LET /IN   construct. The LET
clause consists of a sequence of definitions whose scope extends until the end of the
IN   clause. These local definitions can be used to shorten expressions by replacing
common subexpressions with an operator. It can also make an expression easier to
read. Nested LET clauses are allowed and are a good idea in large specifications.

Also introduced here is the operator CHOOSE. The statement CHOOSE $x : F$
equals an arbitrarily chosen value x that satisfies the formula F. Determinism is
achieved if only one value of x satisfies F.

In *LocateD*, *right* is *chosen to be* the smallest element of *posr*, the set of all
unmarked nodes whose key is greater or equal than the search key; *left* is *chosen to
be* the highest element of *posl*, the set of all unmarked nodes whose key is smaller
than the search key.

Action $CasD2(p)$, in Fig. 3.7, performs the *physical deletion* of the node from
the linked-list. It is guarded by the pre-condition $procDel =$ "swapedD1", which
represents that all the other intermediate actions have already been executed.
It models the CAS operation: if the left node is still pointing to the right one
$mem[AdrLeft].next = AdrRight$, it is made to point to the node identified as
its immediate successor $RigNext$. Otherwise, it is sent back to the location phase

298    $CasD2(p) \triangleq \wedge proc[p].procDel = \text{"swapedD1"}$

300           $\wedge \text{IF } mem[proc[p].ninfo.AdrLeft].next = proc[p].ninfo.AdrRight \text{ THEN}$

302             $\wedge mem' = [mem \text{ EXCEPT } ![proc[p].ninfo.AdrLeft].next = proc[p].ninfo.RigNext]$

304             $\wedge proc' = [proc \text{ EXCEPT } ![p].ninfo.CNkey = 0,$
305                             $![p].ninfo.CNnext = 0,$
306                             $![p].ninfo.AdrLeft = 0,$
307                             $![p].ninfo.AdrRight = 0,$
308                             $![p].ninfo.RigNext = 0,$
309                             $![p].procDel = \text{"readyD"},$
310                             $![p].choice = \text{"undecided"} \ ]$

312             $\wedge \text{UNCHANGED } \langle setup \rangle$

314   ELSE   search again
315             $\wedge proc' = [proc \text{ EXCEPT } ![p].ninfo.AdrLeft = 0,$
316                                 $![p].ninfo.AdrRight = 0,$
317                                 $![p].ninfo.RigNext = 0,$
318                                 $![p].procDel = \text{"identifiedD"}]$
319             $\wedge \text{UNCHANGED } \langle mem, setup \rangle$

Fig. 3.7 Definition of action *CasD2*

($procDel = \text{"identifiedD"}$). Note that there is no *garbage collection*. The memory location is not erased. The is due to the reasons explained in Sect. 3.2.4.

3.4 Promela and SPIN

3.4.1 Overview

Whereas TLA$^+$ was first developed as a specification language on its own, and TLC came after as a model checker that could handle it, Promela is the specifically developed input language of SPIN. Promela resembles an imperative language like C augmented with a few communication primitives. Promela allows the description of the behavior of each process in a system, as well as the interactions between them. For communication, the processes may use FIFO communications channels, rendez-vous or shared variables. SPIN (as TLC) cannot handle infinite state systems, but provides several state space reduction methods, such as, state compression, on-the-fly verification, and hashing techniques. SPIN can check safety and liveness properties of a specification. When a property is violated, SPIN presents a counter-example.

A Promela model is constructed from three basic types of objects:

- Processes
- Data objects
- Message channels

A process is identified by the keyword **proctype** and can have several instantiations. This can be done with the prefix **active** [n], where n is the number of instantiations, or by using the operator **run**[13]. Example:

    active 2 **proctype** ProcessA(parameter1)

The description of a system in Promela starts with constants definition and global variables declarations. Promela is a typed language and all variables must be declared before being referenced. There are two levels of scope: global and process local. The global variables can be accessed by all processes, functioning as the *shared memory* of the system[14].

The types of variables are **bit**, **bool**, **byte**, **chan**, **mtype**, pid, **short**, **int** and **unsigned**.

From the provided basic data types, new ones can be created, in a scalable fashion, through **typedef**.

Message channels are used to model the exchange of data between processes. They are declared through **chan** and can be either local or global, as any other kind of variable. Similarly as in CSP, channame! is used to specify the sending of a message through channel channame, and channame? the reception from the channel. Communication can be asynchronous or "rendezvous". The distinction is made through the size of the buffer of the channel: size zero defines a rendezvous port. A channel declaration has than the form:

    chan name = [buffer size] **of** { variable(s) type}

With **mtype** we can introduce variables that can hold symbolic values defined by the programmer (introduced with one or more **mtype** declarations). A limitation is that no distinction of different sets can be made, i.e., the separate declarations:

```
1  mtype = { data, control, error };
2  mtype = { chair, table, box };
```

are indistinguishable in SPIN from the single declaration:

```
1  mtype = { data, control, error, chair, table, box };
```

A central concept in Promela is **executability**. Every statement in a model is either *executable* or *blocked*, depending on the system state. This provides the basic means for modeling process synchronizations. Print and assignment statements are always executable. Booleans conditions are executable iff[15] they are true. Any statement that is non-executable can block the executing process. Hence, if we want an action to happen only after certain condition(s) we can simply write:

```
1  (conditionA == true);
2  /* description of the action */
```

---

[13]For differentiation each process is assigned a unique instantiation number. This is done automatically, and the user does not need to worry about it. If necessary, each process can refer to its own instantiation number via the predefined local variable _pid.

[14]No intermediate levels can created, i.e. the scope of a global variable cannot be restricted to a subset of processes and the scope of a local one to specific blocks of statements.

[15]*if and only if.*

For control flow, Promela supports the selection statement `if..fi` and the repetition loop `do..od`. Examples:

```
1  if
2  :: (a != b) -> option1
3  :: (a == b) -> option2 fi
4
5  do
6  :: (a > b) -> a = a - b
7  :: (a < b) -> b = b - a
8  :: (a == b) -> break
9  od
```

The element preceding `->` is the guard of the action[16]. Note that the guards need not to be mutually exclusive. If more than one guard is executable, one of the corresponding sequences will be selected nondeterministically.

SPIN can be used to check both safety and liveness properties of a system. The two main constructs to define the **correctness** requirements are *assertions* and *never claims*. An assertion statement has the form `assert (expression)` and is defined locally to a process, therefore checking the correctness of a system in a particular state. To define a *system invariant* – a property that should hold in every reachable state of the system, a `never` claim should be used.[17]

Some properties, like the absence of *deadlock* and *race conditions*, are so basic that they are checked by default. Note that expressing correctness conditions through `never` claims, makes one work with *negated formulas*, i.e. if we want to express that a property p should always hold, what we actually write is that *non p* never occurs. To write a `never` claim that checks for the invariance of the system property p we can write:

```
1  never {
2      do
3      :: !p -> break
4      :: else
5      od
6  }
```

A `never` claim can also be used to express properties in linear temporal logic (LTL). Strictly speaking, Promela does not include syntax for the specification of LTL formulas; SPIN has a separate parser that mechanically translates such formulas into Promela syntax, so that LTL can effectively become part of the language that is accepted by SPIN.

Another construct of Promela to deal with correctness claims are `labels`: By default a valid end state is one in which every process that was instantiated has reached the end of its code. However, if some processes do not reach this point,

---

[16]This notation is inspired in (Dijkstra 1975).

[17]This is not mandatory though, for some properties it can be achieved with an assertion clause through the construction: `proctype` monitor() `assert` (invariant), but this looks a "dirtier" solution.

this may not necessarily be an error.[18] Therefore, to identify individual process states as valid end states the label end can be used. Similarly, there are also the labels progress and accept. The first to *validate* potentially infinite execution cycles, in the case that they visit the labeled state, and the latter to mark states that should not be part of any potentially infinite execution cycle.

```
1   #define L      4      /* Length of the memory */
2   #define M      2      /* Number of inserting processes */
3   #define N      2      /* Number of deleting processes */
4   #define HEAD   1      /* Key value of the 'Head' */
5   #define TAIL   100    /* Key value of the 'Tail' */
6   typedef Node
7   {
8     byte key;
9     byte next;
10    bool mark
11  }
12  Node mem[L];
13  bool setdone=false
```

Listing 3.3 Header of the developed Promela model

3.4.2   Model Developed

The developed Promela model of Harris algorithm is now presented. Design decisions and extracts of the model are explained. The full model is shown in Appendix B. The model was chronologically developed after (and made *as similar as reasonably advisable* to) the one written in TLA⁺. This description contains references to the latter so it should be read after Sect. 3.3.2.[19]

From the three basic Promela types of objects, message channels need not to be used. Three processes and two global variables were used. The processes defined were *setup*, *insert* and *delete*. The last two contain all the operations necessary for the insertion/deletion of a node in/from the list; *setup* is an "auxiliary process" that inserts *Head* and *Tail* in the list.

The *main* global variable of the system is mem, defined as an array of nodes. Node is a user defined type of variable that contains three fields: *key*, *next* and *mark*. setdone is a boolean variable that equals true once *Head* and *Tail* are inserted in the list. Listing 3.3 shows the *header* of the model.

The processes in Promela are defined in a sequential manner. SPIN makes no assumption on the relative speed of processes execution and, when model-checking the system it automatically generates interleaving sequences of execution

---

[18]It can be perfectly valid, for instance, for server processes to enter a wait state after a transition is completed, immune to the fact that all user processes have terminated.

[19]And after Sect. 3.4.1, naturally, if you are not familiar with Promela.

of the processes.[20] Throughout the processes specification there are comment lines referring to the equivalent TLA$^+$ action/textual description phase, e.g. `/* CreateI - Ii */`.[21] The names of the variables are also similar.

The correctness claim of the model – equivalent to *Coherence* shown in (Fig. 3.4, page 57), in TLA$^+$ – expresses that all the nodes registered in mem and that point to another node, have a key that is strictly smaller than the key of the node that they point to. This was expressed in Promela as show in Listing 3.4.[22]

```
1  never
2  {
3    do
4    :: !( !( (mem.key != 0) \&\& (mem.next != 0) ) ||
5       (mem.key < mem[mem.next].key)) -> break
6    :: else
7    od
8  }
```

Listing 3.4   Never claim in the developed Promela model

Explanation:

`(mem.key != 0) && (mem.next != 0)` – nodes registered that point to another node, *elegible*.

`(mem.key < mem[mem.next].key)` – key strictly smaller than the key of the node that they point to, *keysmaller*.

We want to express: *elegible* implies *keysmaller*, *p*. This is logically equivalent to: not *elegible* or *keysmaller*, *p*. The final form comes from Promela syntax.

proctype insert() starts (Listing 3.5) with the statement (setdone==true). This serves as a synchronization method: an inserting processes can only start to execute after the *setup* is done. Till then it is blocked.

The guard (L - nonblank - M) > 0) expresses that an insertion can only take place if the memory is not full. The *generation* of the key to insert[23] is a bit *archaic*. Promela provides no predefined function to generate/select a random number from a set.

The *physical insertion* of the node (Listing 3.6) ends the **proctype**. The node is registered in mem in the lowest available blank position, pos. **atomic** { ... }, both in Listing 3.5 and 3.6, is used to define a fragment of code that is to be executed atomically.

The definition of **proctype** delete() brings no new aspects to this description. It starts with the statement (setdone==true), declaration of the local variables

---

[20] Note that operations like reads and writes are executed atomically.

[21] See also Table 3.1 on page 57.

[22] Recall that a never claim that checks for the invariance of system property *p* can be written in Promela as:

never do :: !p -> **break** :: **else od**

[23] **if** ::CNkey=10 ... ::CNkey=50 **fi**

```
1   active [M] proctype insert()
2   {
3     (setdone==true); /* Guarding condition for the executability of any
4      inserting process -- the initial setup has to be finished first */
5     byte CNkey, CNnext, AdrLeft, AdrRight, t, t_next,  left_next;
6     byte pos=0, counter=0, nonblank=0;
7     startinsert:
8     /*  "CreateI - Ii"  */
9     atomic
10    {
11      do /* count the elements in the list */
12      :: (mem[counter].key != 0 \&\& mem[counter].key < TAIL) ->
13         nonblank ++;
14         counter = mem[counter].next;
15      :: else -> break
16      od;
17      if   /* an insertion can only happen if there's space in mem */
18      :: ((L - nonblank - M) > 0 ) ->
19         if
20         :: CNkey=10
21         :: CNkey=20
22         :: CNkey=30
23         :: CNkey=40
24         :: CNkey=50
25         fi
26      :: else -> goto endinsert
27      fi
28    }
```

Listing 3.5 Beginning of **proctype** insert()

```
1   /* ''CasI2 - Vi'' */
2   atomic
3   {
4     do
5     :: (mem[pos].key != 0) -> pos++
6     :: else -> break
7     od;
8   }
9   if
10  :: (mem[AdrLeft].next == AdrRight) ->
11     mem[AdrLeft].next = pos;
12     mem[pos].key = CNkey;
13     mem[pos].next = CNnext;
14  :: else -> goto searchagain
15  fi;
16  endinsert:
17    skip;
18  } /* End of proctyte insert    */
```

Listing 3.6 End of **proctype** insert()

```
1    /* LocateD - IId */
2    searchagainD:
3       t=0;
4       t_next=mem[0].next;
5       left_next=0;
6       pos=0;
7       do
8       :: ((mem[t_next].mark==true) || (mem[t].key < CNkey)) ->
9          if
10         :: (mem[t_next].mark==false) ->
11            AdrLeft = t;
12            left_next = t_next;
13         :: else ->skip
14         fi;
15         t=t_next;
16         if
17         :: (t==1) -> goto endcycleD
18         :: else ->skip
19         fi;
20         t_next = mem[t].next;
21      :: else -> break
22      od;
23   endcycleD:
24      AdrRight=t;
25      if
26      :: (left_next == AdrRight) ->
27         if
28         :: ((AdrRight!=1) && (mem[mem[AdrRight].next].mark == true))
29            ->
30            goto searchagainD
31         :: else -> goto endsearchD
32         fi
33      :: else -> goto endsearchD
34      fi;
35   endsearchD:
36      skip;
37   /* Examine - IIId */
38      if
39      :: ((AdrRight!=1) || (mem[AdrRight].key != CNkey)) -> goto
40         enddelete
41      :: else -> skip
42      fi;
```

Listing 3.7 Identification of the left and right nodes

and the *archaic generation* of the node to delete. The search of the left and right nodes (Listing 3.7), in line with the others (**proctype** insert() and TLA$^+$ model), implements the simplification stated in the Sect. 3.2.3. The removal of the node from the linked-list is also executed with no *garbage collection*.[24]

---

[24]See Sect. 3.2.4.

3.5   Comparison

Naturally, both formalisms have strengths and weaknesses over each other. To make a comparison as objective as possible, several *measurable* comparison criteria were defined.[25] These criteria were divided into three categories, according to their scope: global, specific for the tool – SPIN or TLC – and specific for the language – Promela or TLA$^+$.

Figure 3.8 summarizes the results of the evaluation made. For every criteria, a *classification* from 0 (minimum) to 4 (maximum) was given.[26] Note that simply adding up the classifications should not be the basis for a decision, some criteria may have a considerably different importance depending on the system to model.

3.5.1   *Matching of the Method to the Application*

As both systems are model checkers they have many aspects in common. They are both appropriate to analyse concurrent systems and both have support for mechanisms as the expression of properties in temporal logic.

3.5.2   *Human Factors*

Compared to other formal approaches that imply the writing of proofs, model checkers can be considered of easier to use. (Even though, the necessity for learning and training is, certainly, not negligible.) Because a significant part of the job is done automatically by the tool, model checkers can be more *productive* than techniques like theorem proving.

3.5.3   *Widespread Utilization*

Model checkers have achieved a relatively good popularity as a technique for the analysis and verification of concurrent systems. TLA$^+$/ TLC has been used in the design and validation of protocols at places like Compaq, Intel and, more recently, Microsoft. SPIN is, however, a more widespread tool, and definitely a very popular one in the field. One factor that could explain this is that TLC considerably younger than SPIN.

---

[25] Several sources of inspiration were used for this exercise; most notably, Sifakis (2010).

[26] In the figure, □ is used if both systems are given an equivalent classification and ◇ otherwise.

COMPARISON CRITERIA			Promela / SPIN	TLA+ / TLC
1. **GLOBAL**	1.1 Matching of the method to the application		□ □ □	□ □ □
	1.2 Human factors	1.2.1 Ease of use	□ □ □	□ □ □
		1.2.2 Productivity	□ □	□ □
	1.3 Widespread utilization		◊ ◊ ◊	◊ ◊
2. **TOOL**	2.1 Licensing / Distribution		□ □ □ □	□ □ □ □
	2.2 Maturity		◊ ◊ ◊	◊
	2.3 Performance	2.3.1 State space	□ □	□ □
		2.3.2 Speed	◊ ◊ ◊	◊
	2.4 Interface		□ □	□ □
	2.5 Coverage of the input language		□ □ □	□ □ □
3. **LANGUAGE**	3.1 Bibliography	3.1.1 Availability	◊ ◊ ◊	◊ ◊ ◊ ◊
		3.1.2 Quality	◊ ◊	◊ ◊ ◊ ◊
	3.2 Expressiveness		◊	◊ ◊ ◊ ◊
	3.3 Readability		◊	◊ ◊ ◊ ◊
	3.4 Reusability		◊	◊ ◊ ◊
	3.5 Scalability / evolutivity		◊	◊ ◊ ◊
	3.6 Level of abstraction		□ □ □	□ □ □
	3.7 Checking possibilities		□ □ □	□ □ □
	3.8 Lifecycle coverage	3.8.1 Requirements	- - - -	- - - -
		3.8.2 Specification	◊ ◊	◊ ◊ ◊ ◊
		3.8.3 Implementation	◊ ◊ ◊	- - - -

Fig. 3.8 Comparison of the formalisms

3.5.4   Licensing/Distribution

Both tools are excellent is this aspect. They require no license fee and can be
freely downloaded from the internet. They can also both run in Windows and Unix
systems. (SPIN also in Mac).

3.5.5   Maturity

Both tools are continuously being upgraded. TLC (from 1999) is, however,
considerably younger than SPIN (from 1991).

3.5.6   Performance

Both tools have similar limitations: they are both explicit state model checkers and
cannot handle infinite systems. All the variables have to be bound. Because TLC
is written in Java it is slower than SPIN, whose verifications are executed with a C
compiled file.

3.5.7   Interface

The interfaces of both SPIN and TLC are generally simple. They both run in
batch mode and, when finding an error, produce a counter example reasonably
understandable. SPIN has a very simple graphic interface.

3.5.8   Coverage of the Input Language

Promela is the specifically developed input language of SPIN. Therefore, it is fully
covered by it. On the other hand, TLA⁺ was first developed than before TLC, as a
high-level specification language. Because TLA⁺ has such high-level capabilities,
TLC cannot handle all of it. This is not a practical problem, because TLC handles
all the specifications that "arise in describing actual systems" (Lamport 2002a).

3.5.9 · Bibliography

Both languages have a reference manual, written by their developers: (Holzmann 2003a) for Promela and (Lamport 2002b) for TLA$^+$. Lamport (2002b) can, however, be freely downloaded from the internet.[27]

Lamport (2002b) is also a very well written book. It presents TLA$^+$ with a series of examples growing in complexity and makes a great effort in making the reader understand the principles of the language, enlightening the fact that TLA$^+$ provides a nice way to formalize the style of reasoning about systems. In Roscoe (1998), one can find the quote: "*In order to use these tools effectively you need a good grasp of the fundamentals of CSP: the tools are certainly not an alternative to gaining an understanding of the theory. Therefore, this book is still, in the first instance, a text on the principles of the language rather than being a manual on how to apply its tools.*" Holzmann (2003a) has not so well succeeded in this aspect. Also, several dispensable references to similarities with C are constantly present, as well as "dubious alternative solutions" to problems with a simple clean one.[28]

3.5.10 Expressiveness

TLA$^+$ is clearly more expressive than Promela. It guides us to think logically and reason about the system to develop and everything can be expressed as *what we want* and not on *how to get there*, a capability very useful when specifying systems. Constructions like do .. od loops to compute the value of a variable need not to be used, they can be more simply expressed as a function. Recall, for instance, the use of Cardinality in *CreateI*, the definitions of *left* and *right* in *LocateD*, or the *archaic generation* of a node in proctype insert().

Another aspect is the correctness claims of the models. The expression of *Coherence* – Fig. 3.4, is more natural and it is easier to derive its equivalent never claim in Promela – Listing 3.4. As an *historical note* it can be noted that the latter was considerably harder to reach. In a first approach and *brainwashed* by the C like look of Promela, the result was:

```
1   never
2   {
3      do
4      ::  (mem[index].next != 0) ->
5          if
6          ::  (mem[index].key < mem[mem[index].next].key) ->
7              skip;
8              index = mem[index].next;
```

---

[27] http://research.microsoft.com/users/lamport/tla/book.html

[28] As a curiosity, you can, for instance, see page 51.

```
9      :: else ->
10          index=0;
11          wrong=true;
12          break
13       fi
14    :: wrong -> break
15    :: else
16    od
17  }
```

3.5.11   Readability

A model of reasonable dimension is more easily read in TLA$^+$ than in Promela. Understanding the behavior of a system modeled in TLA$^+$ can be achieved by reading its *header* and then the *bottom* of the module. Furthermore, the automatically generated LaTeX documents of the models give them a nice clean look.

One may refute this idea by aversion to mathematical symbols. This may be natural because of some unfamiliarity and lack of use in dealing with them, but all it takes is that initial step. Below is an example on how powerful it can be.

Take a lights system. One property about them is that *every time that the light is yellow, it eventually turns red*. In an LTL formula this is expressed as:

$$\Box((L = yellow) \Rightarrow \Diamond(L = red))$$

It should take no more than being told that $\Box$ reads *always*, $\Rightarrow$ reads *implies* and $\Diamond$ *eventually*, to understand the previous and similar expressions.

Now, recall that Promela does not include syntax for the specification of LTL formulas – SPIN has a separate parser that mechanically translates such formulas into Promela syntax. Abbreviating the formula as:

$$\Box(p \Rightarrow \Diamond q)$$

and running this parser produces the result:

```
1  never /* [] (p -> <> q) */
2  {
3    T0_init:
4      if
5      :: ((! (p)) || (q)) -> goto accept_S20
6      :: (1) -> goto T0_S27
7      fi;
8    accept_S20:
9      if
10     :: ((! (p)) || (q)) -> goto T0_init
11     :: (1) -> goto T0_S27
12     fi;
13   accept_S27:
14     if
```

```
15      :: (q) -> goto T0_init
16      :: (1) -> goto T0_S27
17      fi;
18   T0_S27:
19      if
20      :: (q) -> goto accept_S20
21      :: (1) -> goto T0_S27
22      :: (q) -> goto accept_S27
23      fi;
24   }
```

free of the LTL operators but *practicably unreadable*. Please note that this is not a criticism to Promela/SPIN – what we want to debate is the aversion that one can have to mathematical symbols. The result shown is fully automatically generated by SPIN, by simply typing: spin -f ''[](p -> <> q)'' in the command line, and the *readable* original LTL formula is shown as a comment (between /* */) at the beginning.

3.5.12 Reusability

Reusability is an issue in computer science. It is achieved by the possibility of creating modular, parametric or generic descriptions. TLA$^+$ incorporates this features significantly better than Promela. A specification is easily spread over different modules, that are then related with the commands EXTENDS and INSTANCE. This is not possible in Promela.

3.5.13 Scalability

This concerns the possibility of modifying or extending descriptions in an incremental manner. It is also related to modular, parametric and generic descriptions.

3.5.14 Level of Abstraction

In this aspect we can say that TLA$^+$ and Promela have *equivalent capabilities*. It is important to be able to specify a system at different levels of abstraction. TLA$^+$ can go to a higher-level than Promela and, in turn, Promela can go to a level closer to the final implementation solution.

3.5.15   Checking Possibilities

Promela and TLA$^+$ have similar capabilities of expressing/checking correctness claims in combination with their tools, SPIN and TLC.

3.5.16   Coverage of the Lifecycle

None of the languages is appropriate to expressing requirements. TLA$^+$ is a specification language of excellence. The proximity of a language to the final implementation may be a very important aspect, depending on the use that we intend it for. Promela is clearly superior to TLA$^+$ in this aspect. TLA$^+$ is more suitable for reasoning about protocols, whereas a Promela model can be *close* to the final implementation solution. But as we have witnessed in the project, this results in a tendency to express the models in the implementation language with all its semantic issues. Hence, TLA$^+$ is better as it helps the user to think in a more abstract and implementation independent way.

3.5.16.1   Final Remarks

Following the analysis, an important weighting factor that led to the final choice of TLA$^+$/TLC was the higher importance given to the modelling languages than to the tools. More specifically, to the well-matched characteristics of the language with the object of the modelling, than to the performance of the tools, since the later were not *unreasonably different*. Because TLA$^+$ is a high-level very expressive language it was considered the most appropriate solution to start developing and reasoning about the OpenComRTOS functionality.

Despite the effort in presenting a thorough evaluation, it should again be emphasized that the presented results can still be considered subjective and that they represent no more than the authors' opinion. The successful results of the project seem to corroborate the decision taken. Nevertheless, had the decision been different, the formalization and verification capabilities of any other formalism with would certainly have provided extremely valuable inputs to the project.

Chapter 4
Basic Formal Specification in TLA$^+$

4.1 Introduction

4.1.1 Goal: Awareness in Specifying Systems

Model checking (Clarke et al. 1999; Holzmann 2003b; Lamport 2002a) is a flexible and powerful technique for verifying systems formally with automated tools. In principle, it makes this task easy for the user: given a formal system description and certain conditions to be satisfied, the automated checker either confirms that the system satisfies the conditions, or provides a counterexample.

However, this simplicity of use is deceptive. Stating the system description as well as the conditions requires insight and aptitudes of an essentially mathematical nature. Habrias and Faucou (2004) rightly warn that "'Click and prove' without awareness is nothing but the ruin of formal specifications". Not only are there many pitfalls, but "hiding the math" also deprives the designer of one of the most powerful intellectual tools available. Indeed, the user of automated model or proof checkers can benefit from mathematical reasoning about both the system descriptions (e.g., programs) and the conditions (e.g., invariants, temporal formulas) in various respects. A detailed discussion of these issues is postponed to Appendix B.

For the reasons outlined above, rather than "hiding the math", we aim at making the math very accessible, bringing the benefits within easier reach of the specifier/designer.

4.1.2 A Two-Step Approach

This chapter provides an introduction-by-example to TLA$^+$, illustrating how basic mathematics constitutes a flexible specification language. Examples are taken from

E. Verhulst et al., *Formal Development of a Network-Centric RTOS: Software Engineering for Reliable Embedded Systems*, DOI 10.1007/978-1-4419-9736-4_4, © Springer Science+Business Media, LLC 2011

the OpenComRTOS project. It is meant for designers interested mainly in *reading* specifications; those who also want to *build* them will find information towards the next step in Appendix B.

A very gradual introduction to TLA$^+$ is found in Lamport's book *Specifying Systems* (Lamport 2002a). Here, we follow a slightly steeper path, taking examples directly from the intended application domain. Still, most of this material requires only high school mathematics and ample comments are given, so some *reading* abilities for TLA$^+$ should be obtained by the end of this chapter. For the same reason, the reader need not worry about the more detailed *syntax* of TLA$^+$ in this chapter, as it is introduced in Sect. B.2 by embedding TLA$^+$ in a unifying formalism.

4.2   Structure of TLA$^+$ Specifications

Some insight in the overall structure of a TLA$^+$ specification may help understanding by indicating how to read them.

4.2.1   Basic Structure

A TLA$^+$ specification describes the behaviour of a system in terms of changes of its internal state. This state is by definition the value of the declared variables. A declaration has the form

$$\text{VARIABLES } var_0, \ldots, var_{n-1}.$$

Variables may denote entities of any complexity, from simple booleans to functions (including records and sequences).

State changes are expressed by means of *guarded actions*.

A *guard* is a proposition about the current state that expresses when the action can take place. Propositions about the current state contain only variables without prime ($'$).

An *action* is a proposition where variables may occur primed, in which case they denote components of the next state. For instance, $x' = x + 1$ defines the new value x' to be $x + 1$. Two brief asides.

- When writing $x + 1$ one typically assumes that x is a number; otherwise $x + 1$ can be anything. This issue will be taken up again later.
- An action must involve all state variables; those that do not change must be captured by a proposition of the form UNCHANGED $\langle var_{k_0}, \ldots, var_{k_{m-1}} \rangle$ with obvious meaning ($var'_{k_i} = var_{k_i}$ for all i in $0 .. m - 1$).

A specification can contain several guarded actions in the following form.

$$A_0 \stackrel{\Delta}{=} guard_0 \wedge action_0$$

$$\dots$$

$$A_{j-1} \stackrel{\Delta}{=} guard_{j-1} \wedge action_{j-1}$$

$$Next \stackrel{\Delta}{=} A_0 \vee \dots \vee A_{j-1}.$$

where each $action_i$ may itself be fairly complex. In the overall action *Next*, an action A_i that holds for the current state s (abbreviating the tuple $\langle var_0, \dots, var_{n-1} \rangle$) and at least some value s' (abbreviating $\langle var'_0, \dots, var'_{n-1} \rangle$) for the next state is said to be *enabled*. *Next* holds for specific s and s' iff at least one of the A_i holds for that s and s'. The guards can be used for synchronization in an interleaving model of concurrency.

To allow concurrent operation with other modules without forcing a *Next* step at every global state change, TLA⁺ specifications insert *stuttering steps* by not just using *Next* but rather $[Next]_s$, which stands for $Next \vee s' = s$.

A typical complete module specification has the form

$$Spec \stackrel{\Delta}{=} Init \wedge \Box [Next]_s.$$

Here, *Init* is a proposition characterizing the initial state, and $\Box$, read "always" or "henceforth" is a *temporal operator* specifying that its "argument" $[Next]_s$ must be satisfied at every state transition.

A typical property that one might associate with a module specification is an *invariant*, i.e., a state proposition *Invar* such that

$$Spec \Rightarrow \Box Invar.$$

Here, $\Rightarrow$ is a standard symbol for logical implication. In words, this formula says the following: if the system satisfies *Spec*, then *Invar* always holds (i.e., in every state encountered in the system's behaviour). Intuitively, $\Box Invar$ provided (a) *Init* $\Rightarrow$ *Invar* and (b) in case the current state s satisfies *Invar*, then any next state satisfying $[Next]_s$ satisfies *Invar'* (which denotes *Invar* with all state variables primed).

Example: if *Init* ensures that x is a number (e.g., *Init* $\stackrel{\Delta}{=} x = 3$) and *Next* entails $x' = x + 1$, then x will always be a number. A property such as "always being a number" is called a *type invariant*. Note that TLA⁺ itself is untyped, but one can prove type-related properties of the kind just illustrated.

4.2.2 Module Structure

A *module* is a named entity that declares constants and variables (called the *parameters*) and defines a number of symbols such as the A_i, *Init*, *Next*, *Spec* introduced in Sect. 4.2.1.

Modules can be replicated by *instantiation*, renaming parameters as desired. A given module M can be instantiated within another module by writing

$$I \stackrel{\Delta}{=} \text{INSTANCE } M \text{ WITH } m_0 \leftarrow e_0, \dots, m_{k-1} \leftarrow e_{k-1},$$

where the m_i are parameters declared in M and the e_i have meaning inside the current module (usually also declared parameters). This creates within the current module an instance I of M, which means that every symbol a defined in M is available within the current module under the name $I!a$. The definition of $I!a$ is derived from the definition of a in M by replacing every m_i by e_i.

Conceptually, the list of m_i must contain all parameters of M but, by convention, omitting parameter m_i amounts to writing $m_i \leftarrow m_i$.

There are some variants whose principles can be inferred from their usage, and they are also discussed in (Lamport 2002a).

4.3   Introducing TLA$^+$ By Example

We have chosen examples simple enough to explain in detail, yet sufficiently varied to illustrate the most essential features of TLA$^+$.

4.3.1   Basic TLA$^+$ Notions

In TLA$^+$, most concepts are the same as in mathematics. However, in view of formalizing them sufficiently for providing computer support, some of them require a little explanation and notational conventions.

As in mathematics, a *function* f is a mapping that assigns to every element a in some given set A exactly one element, written $f[a]$ and called the *image* of a under f. The set A is the *domain* of f, written DOMAIN f. The domain and the mapping fully define the function. Note that $f[a]$ is considered an error in case $a \notin$ DOMAIN f.

One writes $[A \rightarrow B]$ for the set of all functions from set A to set B.

The expression $[a \in A \mapsto e]$, where e is any expression, denotes the function that maps every a in A to e (which possibly depends on a). Formally,

DOMAIN $[a \in A \mapsto e] = A$ and $[a \in A \mapsto e][d] = (e$ with d substituted for a)

assuming the value of d is in A. For instance, $[n \in Nat \mapsto n+1]$ maps every natural number to its successor and $[n \in Nat \mapsto n+1][2*m] = 2*m+1$ provided $2*m \in Nat$.

An expression of the form $[a \in A \mapsto e]$ is called a *lambda-expression*.

An *operator* is similar to a function, but different in the sense that an operator does not have a domain. Furthermore, an operator must always occur in the form *oper*(*expr*), and *oper* by itself is illegal. Also, *expr* may be anything (no domain!), but if something nonsensical is written, then the value of *oper*(*expr*) may be anything. Conceptually, TLA$^+$ introduces operators to avoid logical paradoxes (sets that are "too large" in some sense).

In TLA$^+$ there is some preference for operators over functions. However, complex data structures that are to be used as variables must be denoted by functions, since a function is a mathematical concept by itself and, if f is a function, then both f and $f[a]$ are syntactically correct.

4.3.2 Basic Examples: TLA$^+$ Sequences and OpenComRTOS Lists

Lists are an ubiquitous concept in a RTOS, used for queueing in FIFO fashion or in order of priority, or other strategies. They are modelled in TLA$^+$ using the module *Sequences*, which itself constitutes a very good introductory example.

4.3.2.1 The Module *Sequences*

This module defines finite sequences, considered as functions with domain $1 .. n$ for some natural number n. As in most programming languages, $m .. n$ is the set of natural numbers starting with m up to and including n (empty if $n < m$).

Here are the main operators on sequences, expressed formally, in words and with an example.

$Seq(S) \triangleq$ UNION $\{[1 .. n \rightarrow S] : n \in Nat\}$.

　　　　The set of all finite sequences of elements from set S. Example: if
　　　　$S = \{a, b\}$ then $Seq(S) = \{\langle \rangle, \langle a \rangle, \langle b \rangle, \langle a, a \rangle, \langle a, b \rangle, \langle b, a \rangle, \langle b, b \rangle, \ldots \}$.

$Len(s) \triangleq$ CHOOSE $n \in Nat :$ DOMAIN $s = 1 .. n$.

　　　　The length of sequence s. Example: $Len(\langle 2, 6, 4 \rangle) = 3$.

$s \circ t \triangleq [i \in 1 .. (Len(s) + Len(t)) \mapsto$ IF $i \leq Len(s)$ THEN $s[i]$
　　　　　　　　　　　　　　　　　　　　　　ELSE $t[i - Len(s)]]$.

　　　　The concatenation of sequences s and t.
　　　　Example: $\langle 2, 6 \rangle \circ \langle 4, 3 \rangle = \langle 2, 6, 4, 3 \rangle$.

$Head(s) \triangleq s[1]$.　　The first element of s. Example: $Head(\langle 2, 6, 4 \rangle) = 2$.

$Tail(s) \triangleq [i \in 1 .. (Len(s) - 1) \mapsto s[i + 1]]$.　　The rest of s.
　　　　Example: $Tail(\langle 2, 6, 4 \rangle) = \langle 6, 4 \rangle$.

We provide a few annotations on the formal definitions.

For *Seq(S)*. Considering the meaning of $[A \to B]$, clearly $[1 .. n \to S]$ is the set of all sequences of length n consisting of elements of S, assuming $n \in Nat$. Taking the union of all $[1 .. n \to S]$ for natural n yields the set of all finite sequences.

For *Len(s)*. In general, the value of (CHOOSE $x \in X$: *property*) is an element x in set X that satisfies *property*. If several x fit, it is not specified which x is taken (but it is always the same). If no x fit, the value is unspecified. If s is a sequence, exactly one n in *Nat* satisfies DOMAIN $s = 1 .. n$.

For $s \circ t$. The concatenation of sequences s and t is again a sequence; hence, a function. It is specified by a lambda expression, saying that the domain is $1 .. (Len(s) + Len(t))$, and that the i-th element is given by a conditional expression (IF construct), with obvious meaning. Indeed, if the index $i \le Len(s)$, then the i-th element of $s \circ t$ is $s[i]$. If $i > Len(s)$, then the i-th element of $s \circ t$ is $t[i - Len(s)]$.

   Observe that this module does not introduce variables, since it does not describe a system but introduces only mathematical functions to be used elsewhere.

   A technicality: as defined in Lamport's textbook (Lamport 2002a), the script of this module starts with LOCAL INSTANCE *Naturals*, thereby instantiating the module *Naturals* locally, that is: without exporting any elements from *Naturals*. Users of the module *Sequences* also needing *Naturals* must incorporate it explicitly.

4.3.2.2 The OpenComRTOS Module *Lists*

This very small module collects some auxiliary definitions. It starts with the text EXTENDS *Naturals, Sequences* to incorporate these modules, and subsequently declares CONSTANTS *PacketData*, the set of values allowed in the fields of a packet (introduced later).

   The formal parameter in the operator definitions is arbitrarily named *wl*, coming from "waiting list", but this is mathematically insignificant.

$$List\_isEmpty(wl) \overset{\Delta}{=} wl = \langle\rangle$$
$$List\_HeadElement(wl) \overset{\Delta}{=} Head(wl)$$
$$List\_deleteHeadElement(wl) \overset{\Delta}{=} [i \in 1 .. (Len(wl) - 1) \mapsto wl[i+1]]$$
$$List\_deletePacket(wl, index) \overset{\Delta}{=} [i \in 1 .. (Len(wl) - 1) \mapsto \text{IF } i < index \text{ THEN } wl[i]$$
$$\text{ELSE } wl[i+1]]$$
$$NoData \overset{\Delta}{=} \text{CHOOSE } nd : nd \notin PacketData$$
$$EmptyPacket \overset{\Delta}{=} [type \mapsto NoData, RequestingTaskID \mapsto NoData,$$
$$prio \mapsto NoData, destination \mapsto NoData,$$
$$data \mapsto NoData]$$

Compared to the module *Sequences*, the only new element is the use of a *record* structure, namely for *EmptyPacket*. It defines *EmptyPacket* to be a function with domain

$$\{\text{"type"}, \text{"RequestingTaskID"}, \text{"prio"}, \text{"destination"}, \text{"data"}\}.$$

The strings in such a set are called the *field names* of the record. By way of syntactic sugar, the expression *EmptyPacket*["type"] may be written *EmptyPacket.type* and so on.

In this example, *EmptyPacket* maps all field names to *NoData*.

Packets that are not necessarily empty belong to a set of records that can be specified as follows. This information will be deducible later from the other modules, and can be considered as further clarification for these modules:

$$Packet \triangleq [type : \{\text{"SID\_SendPacket"}, \text{"SID\_ReceivePacket"}\} \cup \{NoData\},$$

$$RequestingTaskID : TaskId \cup \{NoData\},$$

$$prio : Priority \cup \{NoData\},$$

$$destination : PortId \cup \{NoData\},$$

$$data : PacketData \cup \{NoData\}]. \tag{4.1}$$

Here, *TaskId*, *Priority*, *PortId* are sets defined as constants in the *Port* module introduced later. By definition (4.1), each element of the set *Packet* is a function specified as follows. The domain is the set of field names specified in (4.1) to the left of the ":". The function maps every field name to some value in the set to the right of the ":" following that field name. For instance, if $p \in Packet$ then $p.prio \in Priority \cup \{NoData\}$.

4.3.3 An Extended Example: The Module Port

Module *Port* is selected as an example because it consists mainly of the elements common to all modules at the L1 layer of OpenComRTOS. It represents the designer's view of the entity *Port*.

4.3.3.1 Informal Specification

Tasks synchronize and exchange data via *Ports*. Tasks and Ports are designated by their IDs, constituting the sets *TaskId* and *PortId* respectively.

A task needing to synchronize issues a Packet to a designated port and thereby becomes inactive. The Packet can be assigned a *send* or *receive* type. As long as Packets arrived at a Port have the same type, the IDs of the issuing Tasks remain in a *Port waiting list* according to priority. When a Packet of different type arrives, its issuing Task and the Task at the head of the Port waiting list are activated and the data in the data fields of the Packets of these Tasks are interchanged.

Here, it must be noted that a Packet is identified by the ID of the issuing Task, as it is *preallocated* to the Task.

Furthermore, Packets issued by Tasks do not go directly to the designated Ports, but their Task ID's wait according to priority in a *Kernel Port waiting list*.

It will be somewhat surprising how much detail the formal version of this informal specification entails.

4.3.3.2 Basic Data Structures

Here, we provide some information that can be useful as a legend to the module as it is written in TLA$^+$.

The constants of the module *Port* are the sets *TaskId*, *Priority* (numbers), *PortId*.

The variables of the module *Port* are listed in the following table. The type invariant is not in the specification, but is inferred from it. It provides insight in the structure of the values that may be assumed by the variables, and makes reading the specification easier. The shorthands in the first column are for later use.

Shorthand	Complete variable name and type invariant	
RL	*ReadyList*	$\in$ SUBSET *TaskId*
KL	*KernelPortWL*	$\in$ *Seq(TaskId)*
task	*task*	$\in$ [*TaskId* $\to$ [*prio* : *Priority*]]
PP	*PreallocatedPacket*	$\in$ [*TaskId* $\to$ *Packet*]
PL	*PortWL*	$\in$ [*PortId* $\to$ *Seq(TaskId)*]

The variables are put to the following use.

- *ReadyList* is rather a misnomer, as it not a list but just a set of Task IDs, namely, of Tasks that are active. Initially, all Tasks are active.
- *KernelPortWL* is the Kernel waiting list where Tasks that have sent a packet await treatment according to priority.
- Clearly *task* is a function that assigns to every Task ID a record indicating its priority. In particular, if $t \in TaskId$, then $task[t].prio \in Priority$.
- *PreallocatedPacket* assigns to every Task a Packet, which is a record whose contents will be changed as needed during its use.
- *PortWL* assigns to every Port a waiting list, used as explained in the informal specification, and formalized later.

The initial state is specified as follows:

$$Init \overset{\Delta}{=} \wedge ReadyList = TaskId$$
$$\wedge KernelPortWL = \langle\rangle$$
$$\wedge task \in [TaskId \to [prio : Priority]]$$
$$\wedge PreallocatedPacket = [t \in TaskId \mapsto EmptyPacket]$$
$$\wedge PortWL = [p \in PortId \mapsto \langle\rangle].$$

The $\wedge$ in the first line is for layout. The TLA$^+$ layout conventions are rather self-explanatory and can also reduce parentheses.

4.3.3.3 Top-Down Formal Specification

The TLA⁺ syntax requires that the specifications of operators precede their use. This usually means that the details precede the overall picture, and the complete specification is given at the end. For understanding, it is better to start with the overall picture, which is the *Next* action and *Spec*.

$$Next \overset{\Delta}{=} \lor \exists t \in TaskId : \exists p \in PortId : \lor L0\_SendPacket(t,p)$$

$$\lor L0\_ReceivePacket(t,p)$$

$$\lor L0\_sendreceivePacketService$$

$$Spec \overset{\Delta}{=} Init \land \Box[Next]_{\langle task,ReadyList,KernelPortWL,PortWL,PreallocatedPacket\rangle}.$$

The existential quantifier $\exists$ is best understood as a means of expanding $\lor$ over the elements of the finite sets *TaskId* and *PortId*. Here is a small example showing the principle.

$$\exists x \in \{a,b\} : P(x) \equiv P(a) \lor P(b).$$

In the *Next* action, we have two kinds of subactions.

(a) First, *L0_SendPacket*(t,p) describes the issuing of a Packet by Task t destined for Port p. Here is the definition.

$$L0\_SendPacket(t,p) \overset{\Delta}{=} \land t \in ReadyList$$

$$\land \exists pack \in [type : \{\text{``SID\_SendPacket''}\},$$

$$Requesting\_TaskID : \{t\}, prio : \{task[t].prio\},$$

$$destination : \{p\}, data : \{NoData\}] :$$

$$\land L0\_insertPacket(t,pack)$$

$$\land \text{UNCHANGED } \langle PortWL, task\rangle$$

L0_ReceivePacket(t,p) is similar, differing only in the type field. Note that the set of records in this definition contains only one record, as the set corresponding to each field name is a singleton set. Observe how the Packet inherits its priority from the issuing Task, and how the specified destination is filled in.

An alternative, perhaps more straightforward style is using LET as shown next.

$$L0\_SendPacket(t,p)$$

$$\overset{\Delta}{=} \land t \in ReadyList$$

$$\land \text{LET } pack \overset{\Delta}{=} [type \mapsto \text{``SID\_SendPacket''},$$

$$Requesting\_TaskID \mapsto t, prio \mapsto task[t].prio,$$

$$destination \mapsto p, data \mapsto NoData] :$$

IN *L0_insertPacket*(*t, pack*)

∧ UNCHANGED ⟨*PortWL, task*⟩

Anyhow, from the UNCHANGED part we deduce that *L0_insertPacket*(*t, pack*) changes the remaining variables *ReadyList, KernelPortWL* and *Preallocated-Packet*. More specifically, it removes the issuing Task from *ReadyList*, inserts this Task's ID in *KernelPortWL* according to priority, and puts the specified content in that Task's *PreallocatedPacket*.

The formal specification is simple yet illustrative.

L0_insertPacket(*pid, pack*) $\triangleq$

    ∧ *ReadyList'* = *ReadyList*\\{*pid*}

    ∧ *KernelPortWL'* = IF *List_isEmpty*(*KernelPortWL*)

           THEN ⟨*pid*⟩

           ELSE *List_insertPacket*(*KernelPortWL, pid, pack*)

    ∧ *PreallocatedPacket'* = [*PreallocatedPacket* EXCEPT ![*pid*] = *pack*].

As a clarification for *ReadyList*\\{*pid*}, we note that $S \setminus T$ is the set of elements of set S that are not in T.

For given non-empty list *wl* (of Packet identifiers), pPacket identifier *pid* and Packet *pack*, the value of *List_insertPacket*(*wl, pid, pack*) is the list *wl* in which *pid* is inserted according to the priority of *pack*. Recall that the Task identifiers are used as Packet identifiers, considering that each Task has its preallocated Packet. Note how the conditional expression checks for emptyness. More elegant would have been doing this check inside the definition of *List_insertPacket*(*wl, pid, pack*).

A new construct is the expression [*PreallocatedPacket* EXCEPT ![*pid*] = *pack*]. In general, [*f* EXCEPT ![*d*] = *e*] denotes a function equal to *f*, except that it maps the domain value *d* to the value of *e*. The expression shown is an example.

(b) Second, *L0_sendreceivePacketService* specifies the rendezvous behaviour. A Packet (Task) identifier is removed from the head of the *KernelPortWL* and processed as follows according to the *PortWL* of the destination port. If this *PortWL* is empty, the identifier is placed there. Otherwise, if the type of the Packet matches that of the head of *PortWL*, it is inserted according to priority. If the types differ, "rendezvous" takes place: both Tasks involved are activated (placed in *ReadyList* by *L0_MakeTaskReady*) and the data fields

$L0\_sendreceivePacketService \overset{\Delta}{=}$

$\land \neg Empty(KL) \land \lor PP[Head(KL)].type = \text{"SID\_SendPacket"}$
$\qquad\qquad\qquad\quad \lor PP[Head(KL)].type = \text{"SID\_ReceivePacket"}$

$\land \text{ LET } p \overset{\Delta}{=} PP[Head(KL)], d \overset{\Delta}{=} p.dest$

$\quad \text{IN } \land PL' = \text{IF } Empty(PL[d])$

$\qquad\qquad\qquad \text{THEN } [PL \text{ EXCEPT } ![d] = \langle Head(KL)\rangle]$

$\qquad\qquad\qquad \text{ELSE } \text{ IF } PP[Head(PL[d])].type = p.type$

$\qquad\qquad\qquad\qquad \text{THEN } [PL \text{ EXCEPT } ![d] = Ins(PL[d], Head(KL), p)]$

$\qquad\qquad\qquad\qquad \text{THEN } [PL \text{ EXCEPT } ![d] = Tail(PL[d])]$

$\qquad \land KL' = Tail(KL)$

$\qquad \land \text{ IF } Len(PL'[d]) < Len(PL[d])$

$\qquad\quad \text{THEN } \land RL' = RL \cup \{p.ReqID, PP[Head(PL[d])].ReqID\}$

$\qquad\qquad\qquad \land PP' = [PP \text{ EXCEPT } ![Head(KL)].data = PP[Head(PL[d])].data,$

$\qquad\qquad\qquad\qquad\qquad\qquad\quad ![Head(PL[d])].data = PP[Head(KL)].data]$

$\qquad\quad \text{ELSE UNCHANGED } \langle RL, PP \rangle$

$\quad \land \text{UNCHANGED } task$

Fig. 4.1 Rendezvous in SendReceivePacketService

of their *PreallocatedPacket* interchanged. The definition is given in Fig. 4.1, where we use the following abbreviations:

KL	for *KernelPortWL*
PL	for *PortWL*
PP	for *PreallocatedPacket*
Ins	for *List_insertPacket*
Head	for *List_HeadElement*
Tail	for *List_deleteHeadElement*
ReqID	for *RequestingTaskID*

to improve the synoptic view and thereby legibility. Also for legibility, in Fig. 4.1 we wrote the action $RL' = RL \cup \{p.ReqID, PP[Head(PL[d])].ReqID\}$, which is equivalent to the action $L0\_MakeTaskReady(p.ReqID, PP[Head(PL[d])].ReqID)$ actually written in the module *Port*.

4.3.3.4 A Final Detail

One of the low-level operators to mention here is *List_insertPacket*, as it the only nontrivial component of *Port* not yet formally defined in the preceding description. It also illustrates one more TLA$^+$ construct, namely the use of CASE. Informally: for given nonempty list *wl* (of Packet identifiers), Packet identifier *pid* and Packet *pack*, the value of *List_insertPacket(wl,pid,pack)* is the list *wl* in which *pid* is inserted according to the priority of *pack*. Here is the formal description.

$List\_insertPacket(wl,pid,pack) \triangleq$

    CASE *pack.prio* $\leq PP[wl[1]].prio \rightarrow$

        $[i \in 1..(Len(wl)+1) \mapsto$ IF $i = 1$ THEN *pid* ELSE $wl[i-1]]$

    □   *pack.prio* $\geq PP[wl[Len(wl)]].prio \rightarrow$

        $[i \in 1..(Len(wl)+1) \mapsto$ IF $i = (Len(wl)+1)$ THEN *pid* ELSE $wl[i]]$

    □   OTHER $\rightarrow [i \in 1..(Len(wl)+1) \mapsto$

            CASE $i = 1 \rightarrow wl[1]$

            □ $i = Len(wl)+1 \rightarrow wl[i-1]$

            □ OTHER $\rightarrow$ IF *pack.prio* $> PP[wl[i]].prio$

                THEN $wl[i]$

                ELSE IF *pack.prio* $> PP[wl[i-1]].prio$

                THEN *pid*

                ELSE $wl[i-1]]]$

The CASE construct is a conditional expression. The general form is

$$\text{CASE } p_0 \rightarrow e_0 \;\square\; p_1 \rightarrow e_1 \;\square\; \cdots \;\square\; p_{n-1} \rightarrow e_{n-1} \;\square\; \text{OTHER} \rightarrow ee$$

where the various p_i are propositions whereas the various e_i and *ee* are any expressions. The value of the construct is one of the e_i for which p_i holds; if none of the e_i holds, the value is *ee*.

4.3.3.5 Checking Potential Issues in the Module

Here are some typical potential issues that were submitted to the model checker when verifying *Port*. Note that while we speak of "issues", in formal modelling terms what is checked are properties of the system that must hold at all time, and hence called "invariants". They were bundled together (using ∧) to a state proposition somewhat arbitrarily called *TypeInvariant*, and submitted as

$$\text{THEOREM } Spec \Rightarrow \square TypeInvariant.$$

Here are some components of *TypeInvariant*, preceded by their informal statement.

(1) There are never more Tasks in the ready list than in the system.

$$Cardinality(ReadyList) \leq Cardinality(TaskId).$$

(2) There are never more Tasks in a Port's waiting list than in the system.

$$\forall\, p \in PortId : Len(PortWL[p]) \leq Cardinality(TaskId).$$

(3) All Tasks waiting at a Port have the same type.

$$\forall\, p \in PortId : \forall\, i,j \in 1\,..Len(PL[p]) : PP[PL[p][i]].type = PP[PL[p][j]].type.$$

(4) All Tasks at the Kernel Port are valid Tasks.

$$\forall\, i \in 1\,..Len(KernelPortWL) : KernelPortWL[i] \in TaskId.$$

(5) All Tasks at the Kernel Port are sorted by priority. Formally, with shorthands:

$$\forall\, i,j \in 1\,..Len(KL) : (i \leq j) \Rightarrow (PP[KL[i]].prio \leq PP[KL[j]].prio).$$

(6) All Tasks at a *PortWL* are identified with their *TaskId*.

$$\forall\, p \in PortId : \forall\, i \in 1\,..Len(PortWL[p]) : PortWL[p][i] \in TaskId.$$

(7) All Tasks at a Port are sorted by priority. Formally, with shorthands:

$$\forall\, p \in PortId : \forall\, i,j \in 1\,..Len(PL[p]) :$$
$$(i \leq j) \Rightarrow (PP[PL[p][i]].prio \leq PP[PL[p][j]].prio).$$

(8) No Task waiting for a Port can be ready.

$$\forall\, p \in PortId : \forall\, i \in 1\,..Len(PortWL[p]) : PortWL[p][i] \notin ReadyList.$$

(9) No Task on the ready list can be waiting at a Port.

$$\forall\, t \in ReadyList : \forall\, p \in PortId : \forall\, i \in 1\,..Len(PortWL[p]) : PortWL[p][i] \neq t.$$

These are just examples of what kind of properties can be checked and how they can be formulated in TLA^{+}.

4.4 Conclusion

In this chapter, we have achieved two goals:

(a) The primary goal was introducing a sufficiently large and representative subset of TLA^{+} by way of relatively simple examples. The *Port* module gave the opportunity to present the most frequently used constructs of TLA^{+}.

Readers interested in a systematic overview of TLA$^+$ from the language view-point are referred to Appendix B,which also exposes some of the mathematics relevant to the semantics of TLA$^+$ and to writing specifications with temporal operators.

(b) This chapter also gave an introduction to a representative entity of OpenComR-TOS together with the associated key concepts and terminology.

Other OpenComRTOS entities will be introduced and discussed in subsequent chapters.

Part III
OpenComRTOS Design

Chapter 5
Formal Modelling of the RTOS Entities

5.1 Introduction

This chapter describes the formal TLA$^+$ models of the OpenComRTOS Layer 1 Interaction Entities (L1-Entities). The L1-Entities represent the API (Application Programmer's Interface) of OpenComRTOS used by the Task entities to build up the application. There are also so-called Layer 0 entities but these are not accessible to the user. The L1 Entities are also all derived from a common so-called Hub Entity. As we will see later, the L1-Entities represent services the operating system offers to the user. The L1-Entities names represent the type of service they provide. For instance the L1-Entity 'Port' offers a mechanism to exchange data between two Tasks. OpenComRTOS (Version 1.1) offers the following L1-Entities:

- Port – A rather generic Hub Entity to exchange Packets between Tasks.
- Event – A (binary) Event Entity to synchronise a Task through a single Boolean Event with another Task. Typically, also used by a specific hardware peripheral to signal the occurrence of a hardware event to its driver task.
- Semaphore – An Entity used to synchronise tasks based on counting Events.
- Resource – An Entity used to provide exclusive ownership of a logical resource.
- Memory Pool – An Entity providing exclusive ownership of memory blocks of a predefined size.
- Packet Pool – An Entity providing exclusive ownership of Packets.
- FIFO – An Entity used to pass fixed size data in a buffered way between Tasks.

For the construction of high reliable OpenComRTOS applications the correct functionality of the L1-Entities is crucial. This is one of the reasons why we used formal modelling to derive the specifications for the L1-Entities before implementing them. A specification describes the properties of a system. Deriving a correct and complete specification is a difficult task which may take as long or even longer than the actual implementation. The process of deriving a specification for a

E. Verhulst et al., *Formal Development of a Network-Centric RTOS: Software Engineering for Reliable Embedded Systems*, DOI 10.1007/978-1-4419-9736-4_5,
© Springer Science+Business Media, LLC 2011

system from its requirements leads to better understanding of the system. After this effort the implementation becomes in many cases trivial. Therefore, it is a good idea to always derive a specification of a system before implementing it!

Formal models can support checking the correct operation and use of an entity only at a specific level of abstraction. Therefore, a formal model may not hold when applied at a different level of abstraction. For example, the models of OpenComR-TOS L1-Entities are not written in and not meant to be C code. Any implementation language can be used, although C is most often the one used for embedded systems. The reason for this is that thinking immediately in an implementation language like C restricts the thinking and will most likely not result in new insights about the system to be specified. To phrase it more directly, when modelling the specification with C one is not deriving a specification but is already implementing and the exercise looses its purpose. The models are only specifications which allow us to check the properties of the system. Nevertheless, one of the modelling goals was to be as close as possible to the implementation. Only a high degree of similarity between the simulation model and the implementation allowed us to transfer the proven properties of the formal model to the system implementation.

To discuss the complete TLA$^+$ model of all OpenComRTOS L1-Entities in detail would exceed the scope of this chapter. Therefore, we concentrate on the TLA$^+$ model of the Semaphore-Entity. We chose the Semaphore as example, because it is a well known synchronisation primitive with no extra functionality. Furthermore, the complexity of the model is such that some finer points of the modelling process can be shown without occupying too much space in this book.

The discussion starts by introducing the base concepts used throughout the TLA$^+$ models of the L1-Entities. Together with the list model given previously (Sect. 4.3.2), these concepts build the foundation to understand the Semaphore model discussed in Sect. 5.3. The model is introduced by stating constraints, model functionality and proof obligations. After this introductory background, the TLA$^+$ model is discussed in detail. The chapter ends with some conclusions.

5.2   OpenComRTOS Environment Model

This section introduces functions and variables which represent the environment of the L1-Entities models. These functions and variables therefore represent the OpenComRTOS core.

It is impossible to model the complete OpenComRTOS system with a single model, because the complexity is too high for a meaningful model. We consider a model to be meaningful if the proven model properties also hold for the implementation. To come around this problem, we used the divide and conquer approach and split the system into smaller, more or less orthogonal parts which could be modelled. However, the drawback of this method is that the state system as a whole is not reflected in the models. To overcome this, we modelled the system state using TLA$^+$ actions and variables. This approach also simplifies the development of the models for the individual L1-Entities, because it provides the templates to develop the specific models while promoting reuse.

5.2.1   Term Definitions

This section discusses how the OpenComRTOS entities are expressed in TLA$^+$. First, we define the base terms of the L1 TLA$^+$ models. These terms describe how OpenComRTOS entities are mapped onto TLA$^+$ expressions. The following list defines the terms used in the TLA$^+$ models of the L1-Entities:

- Task – Entity which is represented as a record with a unique ID.
- Packet – Entity which is used to model interactions between Tasks and Kernel services. Section 5.2.4 gives the details of the L1-Packet.
- List – Is a sequence of packet IDs. Each preallocated Packet has the same ID as the Task to which it belongs. Due to the high usage of Lists throughout the OpenComRTOS Kernel the model of this element will be detailed in Sect. 4.3.2.
- Kernel (Task) – Represented by actions and system variables (Kernel Port WL, etc.).
- Waiting List (further called WL) – List sorted according to priority, highest priority first. If a Packet is in a WL, then the Task, which has sent it, is waiting, as the Packet acts like a placeholder for the Task's state.
- Ready List – A set of TaskIDs. If the TaskID of a Task is in a Ready List then this Task is active[1].

5.2.2   Constants

The OpenComRTOS TLA$^+$ models make use of the following constants:

- *Priority* – set of natural numbers. The priority field of each Task is assigned a value from this set.
- *TaskId* – set of all Task IDs. The cardinality of this set is equal to the amount of Tasks in the system.
- *PacketData* – a set which represents the data part of a Packet. When a data transfer occurs then a packet takes a value from this set.

---

[1]This is called a Ready List because the implementation will use a priority ordered list, but for the formal model, the behavior is priority independent, hence Set is sufficient reducing the state space.

- *NoData* – empty element. It is used when the Packet carries no valid data or when the Packet is used without data transfer.

5.2.3   Variables Representing the System State

The system variables represent the system state. The following list describes the System variables in more detail:

- *task* – record. At this stage of the model development, it consists only of a field named priority.
- *ReadyList* – set of active tasks. Contains the IDs of all active tasks.
- *KernelPortWL* – List which contains the Ids of all waiting tasks that sent a request to a Kernel Port.
- *PreallocatedPacket* – Packet function. The domain of this function is the set of all task IDs, i.e. the Ids of all tasks in the system (constant TaskId). The range of this function is the set of packets. Each task has one Preallocated Packet. The ID of the Preallocated Packet is equal to the Task ID, that owns it.

5.2.4   The L1-Packet

OpenComRTOS uses L1-Packets throughout the system to communicate between different entities. All interactions are L1-Packet exchanges. An L1-Packet consists of these fields:

- *type* – Packet type, can be one of the these three values:
    - *SID_SendPacket* – marks this packet as one which is sent, its content will be copied to the corresponding packet of type *SID_ReceivePacket*.
    - *SID_ReceivePacket* – this packet will get the data from the packet with the type *SID_SendPacket*.
    - *NoData* – this means that the Packet is currently not in use.
- *RequestingTaskID* – the ID of the task which has sent a request to an L1-Entity. RequestingTaskID is an element of the TaskID set.
- *prio* – the priority of the task which has sent a packet. Field prio is an element of the Priority set.
- *destination* – the ID of destination L1 Entity.
- *data* – contains the data to be transferred. This field can be empty when it is not necessary for the desired interaction. For instance to signal a Semaphore (increment its count) all the Semaphore Entity needs to know is that this packet represents a request to increment its count. This is encoded by setting the type of the packet to *SID_SendPacket*.

5.2.4.1   Sending and Receiving Packets

OpenComRTOS Tasks communicate with the L1-Entities by exchanging L1-Packets with them. There are two types of requests a task can send to an L1-Entity: SendPacket or ReceivePacket, which represent the necessary symmetry of actions. In the TLA$^+$ models, these two types of requests are represented by the user actions: $L1\_SendPacket(t,p)$, $L1\_ReceivePacket(t,p)$. Where t is the ID of the Task and p is the ID of the Entity. These actions generate a Packet (send request or receive request) and input them into the Kernel InputPort WL.

5.2.4.2   Making a Task Ready to Run

This function marks either one or two tasks as being ready to run, i.e the Kernel is now able to schedule them:

$$L1\_MakeTaskReady(t1, t2) \;\triangleq$$
$$\wedge ReadyList' = \text{IF } t2 = NoData$$
$$\text{THEN } ReadyList \cup t1$$
$$\text{ELSE } ReadyList \cup t1, t2 \qquad (5.1)$$

5.2.5   General Constraint for All Models

A system based on OpenComRTOS always has at least one active task. Therefore it was necessary to add a constraint which ensures that this applies to the models as well. In essence this constraint avoids the occurrence of a deadlock. Such a deadlock condition can be present in all models, because the checker generates all possible variants and we can have the situation that all tasks sent a request of the same type. The problem is that in order to synchronize and make further progress a pair, i.e. two Packets of complementary type are needed. In the situation, described earlier it is impossible to find a matching pair, therefore no synchronization happens and no further progress is possible, a classical deadlock. At the application level it is the responsibility of the developer to assure that deadlocks cannot happen.

5.3   Formal Model of the Semaphore-Entity

This section details the formal modelling process using the Semaphore-Entity as example.

5.3.1   *Constants*

1. *SemaphoreId* – this is a set which contains the IDs of all L1-Semaphores in the system. The cardinality of this set is the amount of L1-Semaphores in the system.
2. System constants, as defined in Sect. 5.2.2.

5.3.2   *Variables*

1. *SemaphoreWL* – this is a list containing the Packets which contain requests for this specific Semaphore-Entity. The Tasks which sent these request packets are not ready to run.
2. *Count* – this is an array which contains the count of each Semaphore defined in the set *SemaphoreId*.
3. System variables, as defined in Sect. 5.2.3.

The following services are used to interact with a Semaphore:

1. *L1_SignalSemaService* – this service increments the Semaphore Count
2. *L1_TestSemaService* – this service decrements the Semaphore Count.

5.3.3   *Initialisation*

The initialisation should bring the model into the following state:

- All task IDs are in the Ready List (that means all tasks are in thee active state). This implies, all tasks are ready to run.
- All Preallocated Packets are empty.
- All Waiting Lists are empty.
- The Semaphore Count of each Semaphore is zero.

The following TLA statement establishes this initialisation:

$$
\begin{aligned}
Init \overset{\Delta}{=}\ &\wedge ReadyList = TaskId \\
&\wedge task \in [TaskId \rightarrow [prio : Priority]] \\
&\wedge PreallocatedPacket = [t \in TaskId \mapsto EmptyPacket] \\
&\wedge KernelPortWL = \langle\rangle \\
&\wedge SemaphoreWL = [e \in SeamphoreID \mapsto \langle\rangle] \\
&\wedge Count = [e \in SeamphoreID \mapsto 0]
\end{aligned}
\tag{5.2}
$$

After initialization each task can send either SID_SendPacket or SID_ReceivePacket to the Kernel-Input-Port.

5.3.4   Signalling the Semaphore

If a task sent a request of type SID_SendPacket then this results in the invocation of *L1_SignalSemaService*. The specification of *L1_SignalSemaService* in plain English:

- If the Semaphore Waiting list is empty the Semaphore count is incremented. The requesting Task is put on the Ready List again.
- If the Semaphore Waiting List contains a Packet of type SID_ReceivePacket, i.e. a Task is waiting to be signalled from this Semaphore, then we fulfilled the synchronisation predicate. Both Tasks become active again, i.e. both Tasks get inserted into the Ready List. The Semaphore Count does not get modified.

The previously given specification results in the following TLA statements, representing *L1_SignalSemaService*.

The first condition which must be fulfilled in order for the service to be performed is that there is a Packet in the Kernel Port Waiting List:

$$\neg List\_isEmpty(KernelPortWL) \tag{5.3}$$

The second precondition is that this packet must be of type "SID_SendPacket". As all packets in the model are stored in the array *PreallocatedPaket*, each task has a packet in this array, the ID of the task acting as index. All Waiting Lists just store the TaskID instead of a packet itself. The resulting statement is therefore:

$$PreallocatedPacket[List\_HeadElement(KernelPortWL)].type =$$

$$\text{``SID\_SendPacket''} \tag{5.4}$$

Next the model determines the number of times the Semaphore has been signalled after completing this operation (*Count'*). If the Semaphore Waiting List (*SemaphoreWL*) of this Semaphore does not contain any entries then *Count* is incremented by one (*Count'* = @ + 1), otherwise *Count* stays the same.

$$Count' = \text{IF } List\_isEmpty(SemaphoreWL[$$

$$PreallocatedPacket[List\_HeadElement($$

$$KernelPortWL)].destination])$$

$$\text{THEN } [Count \text{ EXCEPT}$$

$$![PreallocatedPacket[List\_HeadElement($$

$$KernelPortWL)].destination] = @ + 1]$$

$$\text{ELSE } Count \tag{5.5}$$

This statement calculates the entries of the Semaphore Waiting List after the operation has completed:

If the Semaphore Waiting List (*SemaphoreWL*) is empty, all that happens is that *Count* is incremented by one. If there are entries in the *SemaphoreWL*, which are always packets which try to test the semaphore, the first element of the list is removed from the *SemaphoreWL* using the function *List_deleteHeadElement*(), because subsequently the corresponding Task will be marked are runnable again:

$$\wedge SemaphoreWL' = \text{IF } List\_isEmpty(SemaphoreWL[PreallocatedPacket[$$

$$List\_HeadElement(KernelPortWL)].destination])$$

$$\text{THEN } SemaphoreWL$$

$$\text{ELSE } [SemaphoreWL \text{ EXCEPT}$$

$$![PreallocatedPacket[List\_HeadElement(KernelPortWL)].$$

$$destination] = List\_deleteHeadElement($$

$$SemaphoreWL[PreallocatedPacket[List\_HeadElement($$

$$KernelPortWL)].destination])] \tag{5.6}$$

After this operation completes the currently handled packet is removed from the Kernel Port Waiting List (*KernelPortWL*):

$$KernelPortWL' = List\_deleteHeadElement(KernelPortWL) \tag{5.7}$$

The Task that sent the Packet signalling the Semaphore is always ready to run after this service has completed:

$$L0\_MakeTaskReady(PreallocatedPacket[List\_HeadElement($$

$$KernelPortWL)].RequestingTaskID, NoData) \tag{5.8}$$

The service does not change the contents of the arrays, *PreallocatedPacket* and *task*:

$$\text{UNCHANGED}\langle PreallocatedPacket, task\rangle \tag{5.9}$$

Equation (5.10) on page 97 is the uncommented version of the model.

$$L1\_SignalSemaService \overset{\Delta}{=}$$

$$\wedge \neg List\_isEmpty(KernelPortWL)$$

$$\wedge PreallocatedPacket[List\_HeadElement(KernelPortWL)].type =$$

$$\text{``SID\_SendPacket''}$$

$$\wedge Count' = \text{IF } List\_isEmpty(SemaphoreWL[$$

$$PreallocatedPacket[List\_HeadElement($$
$$KernelPortWL)].destination])$$

THEN $[Count$ EXCEPT

$$![PreallocatedPacket[List\_HeadElement($$
$$KernelPortWL)].destination] = @ + 1]$$

ELSE $Count$

$\land SemaphoreWL' =$ IF $List\_isEmpty(SemaphoreWL[PreallocatedPacket[$

$List\_HeadElement(KernelPortWL)].destination])$

THEN $SemaphoreWL$

ELSE $[SemaphoreWL$ EXCEPT

$$![PreallocatedPacket[List\_HeadElement(KernelPortWL)].$$
$$destination] = List\_deleteHeadElement($$
$$SemaphoreWL[PreallocatedPacket[List\_HeadElement($$
$$KernelPortWL)].destination])]$$

$\land KernelPortWL' = List\_deleteHeadElement(KernelPortWL)$

$\land L0\_MakeTaskReady(PreallocatedPacket[List\_HeadElement($

$KernelPortWL)].RequestingTaskID, NoData)$

$\land$ UNCHANGED$\langle PreallocatedPacket, task\rangle$ (5.10)

5.3.5   Testing the Semaphore

If a Task sent a request of type SID_ReceivePacket the operation $L1\_TestSema\text{-}$ $Service$ is invoked. Its specification in plain English is:

- If the Semaphore Count is larger than zero, the Semaphore Count is decremented and the requesting Task is put into the Ready List.
- If the Semaphore Count is zero the Packet is inserted into the Semaphore Waiting List according to its priority and the Semaphore Count is not modified.

The previous given specification results in the following TLA statements, which represent the $L1\_TestSemaService$.

One condition which must be fulfilled in order for the service to be performed is that there is a Packet in the Kernel Port Waiting List:

$$\neg List\_isEmpty(KernelPortWL) \qquad (5.11)$$

The second precondition is that this packet must be of type "SID_ReceivePacket". As all packets in the model are stored in the array $PreallocatedPaket$, each task has a packet in this array, the ID of the task acting as index. All Waiting Lists just store the TaskID instead of a packet itself. The resulting statement is therefore:

$$PreallocatedPacket[List\_HeadElement(KernelPortWL)].type =$$

$$\text{"SID\_ReceivePacket"} \tag{5.12}$$

The number of times the Semaphore has been signalled may have to be adjusted once this operation completes. If $Count$ has a value larger than 0 then the Semaphore is already signalled, and it is necessary to decrement the value of count: $Count' = @ - 1$. Otherwise, the value of $Count$ stays the same.

$$Count' = \text{IF } Count[PreallocatedPacket[ListHeadElement($$

$$KernelPortWL)].destination] > 0$$

$$\text{THEN } [Count \text{ EXCEPT!}[PreallocatedPacket[List\_HeadElement($$

$$KernelPortWL)].destination] = @ - 1]$$

$$\text{ELSE } Count \tag{5.13}$$

This statement counts the entries in the Semaphore Waiting ($SemaphoreWL$) List after the operation has completed. If the Semaphore is already signalled, i.e. has a $Count > 0$ then the packet does not have to be inserted into the Semaphore Waiting List. Otherwise, the packet will be inserted into the Kernel Port Waiting List ($KernelPortWL$).

$$SemaphoreWL' =$$

$$\text{IF } Count[PreallocatedPacket[List\_HeadElement(KernelPortWL)].$$

$$destination] > 0$$

$$\text{THEN}; SemaphoreWL$$

$$\text{ELSE IF } List\_isEmpty(SemaphoreWL[PreallocatedPacket[$$

$$List\_HeadElement(KernelPortWL)].destination])$$

$$\text{THEN } [SemaphoreWL \text{ EXCEPT }![PreallocatedPacket[$$

$$List\_HeadElement(KernelPortWL)].destination] =$$

$$\langle List\_HeadElement(KernelPortWL)\rangle]$$

$$\text{ELSE } [SemaphoreWL \text{ EXCEPT }![PreallocatedPacket[$$

$$List\_HeadElement(KernelPortWL)].destination] =$$

$$List\_insertPacket(SemaphoreWL[PreallocatedPacket[$$

$$List\_HeadElement(KernelPortWL).destination],$$
$$List\_HeadElement(KernelPortWL),$$
$$PreallocatedPacket[$$
$$List\_HeadElement(KernelPortWL)])] \quad (5.14)$$

If after this operation the length of the Semaphore Waiting List ($SemaphoreWL'$) has been reduced, then make the Task ready which sent the Packet that got taken off the $SemaphoreWL$. Otherwise, mark $ReadyList$ as being unchanged.

IF $Len(SemaphoreWL'[PreallocatedPacket[List\_HeadElement($

$KernelPortWL)].destination]) < Len(SemaphoreWL[$

$PreallocatedPacket[List\_HeadElement(KernelPortWL)].destination])$

THEN $L0\_MakeTaskReady(PreallocatedPacket[List\_HeadElement($

$KernelPortWL)].RequestingTaskID, NoData)$

ELSE UNCHANGED $ReadyList$ $\quad (5.15)$

After this operation completes the currently handled Packet is removed from the Kernel Port Waiting List ($KernelPortWL$):

$$KernelPortWL' = List\_deleteHeadElement(KenrelPortWL) \quad (5.16)$$

Identify that this operation modifies neither $PreallocatedPackets$ nor $task$.

$$\text{UNCHANGED}\langle PreallocatedPacket, task\rangle \quad (5.17)$$

Equation (5.18) on page 100 is the uncommented version of the model.

$L1\_TestSemaService \overset{\Delta}{=}$

$\quad \wedge \neg List\_isEmpty(KernelPortWL)$

$\quad \wedge PreallocatedPacket[List\_HeadElement(KernelPortWL)].type =$

$\quad \quad$ "SID_ReceivePacket"

$\quad \wedge SemaphoreWL' =$

$\quad \quad$ IF $Count[PreallocatedPacket[List\_HeadElement(KernelPortWL)].$

$\quad \quad \quad destination] > 0$

$\quad \quad$ THEN $SemaphoreWL$

$\quad \quad$ ELSE IF $List\_isEmpty(SemaphoreWL[PreallocatedPacket[$

$\quad \quad \quad \quad List\_HeadElement(KernelPortWL)].destination])$

$\quad \quad \quad$ THEN$[SemaphoreWL$ EXCEPT $![PreallocatedPacket[$

$$List\_HeadElement(KernelPortWL)].destination] =$$

$$\langle List\_HeadElement(KernelPortWL)\rangle]$$

$$\text{ELSE}[SemaphoreWL \text{ EXCEPT } ![PreallocatedPacket[$$

$$List\_HeadElement(KernelPortWL)].destination] =$$

$$List\_insertPacket(SemaphoreWL[PreallocatedPacket[$$

$$List\_HeadElement(KernelPortWL).destination],$$

$$List\_HeadElement(KernelPortWL),$$

$$PreallocatedPacket[List\_HeadElement(KernelPortWL)])])]$$

$$\wedge Count' = \text{IF } Count[PreallocatedPacket[ListHeadElement($$

$$KernelPortWL)].destination] > 0$$

$$\text{THEN } [Count\text{EXCEPT}![PreallocatedPacket[List\_HeadElement($$

$$KernelPortWL)].destination] = @ - 1]$$

$$\text{ELSE } Count$$

$$\wedge KernelPortWL' = List\_deleteHeadElement(KenrelPortWL)$$

$$\wedge \text{IF } Len(SemaphoreWL'[PreallocatedPacket[List\_HeadElement($$

$$KernelPortWL)].destination]) < Len(SemaphoreWL[$$

$$PreallocatedPacket[List\_HeadElement(KernelPortWL)].destination])$$

$$\text{THEN } L0\_MakeTaskReady(PreallocatedPacket[List\_HeadElement($$

$$KernelPortWL)].RequestingTaskID, NoData)$$

$$\text{ELSE UNCHANGED} ReadyList$$

$$\wedge \text{UNCHANGED} \langle PreallocatedPacket, task\rangle \qquad\qquad (5.18)$$

5.3.6   Constraints

There are two constraints placed onto the Semaphore Model. The first constraint is that there always must be at least one entry on the ReadyList. This guards against the case that all tasks decided to Test a semaphore and thus the system cannot make any progress, i.e. deadlocks. In real life it is the task of the developer to ensure this. The second constraint is that a Semaphore is not signalled more than $Limit$ times, this is employed to ensure that the state space does not explode, because without this the maximum value of count is ∞.

$$Constr \overset{\Delta}{=} \wedge Cardinality(ReadyList) > 0$$

$$\wedge \forall s \in SemaphoreId : Count[s] < Limit \qquad (5.19)$$

The TLA$^+$ syntax requires that the specifications of operators precede their use. This usually means that the details precede the overall picture, and the complete specification is given at the end. For understanding, it is better to start with the overall picture, which is the *Next* action and *Spec*.

5.3.7 Defining the Next State

TLA executes the operation called *Next* to determine the next state of the model. Here the previously defined actions get combined:

$$Next \stackrel{\Delta}{=} \vee \exists t \in tId : \exists p \in SemaphoreId :$$

$$\vee L1\_SendPacket(t, p)$$

$$\vee L1\_ReceivePacket(t, p)$$

$$\vee L1\_SignalSemaService$$

$$\vee L1\_TestSemaService \qquad (5.20)$$

$$Spec \stackrel{\Delta}{=} Init \wedge \Box \, [Next] \left\langle \begin{matrix} task, ReadyList, Count, KernelPortWL, \\ Semaphore\,WL, PreallocatedPacket \end{matrix} \right\rangle \qquad (5.21)$$

5.3.8 Properties to Check

This section lists the (invariant) properties which the model of the Semaphore must have. In Sect. 5.3.9 these are translated to TLA proof obligations. Some of the invariant properties checked for are:

1. There are never more task IDs on the ready list than there are Tasks in the system.
2. There are never more task IDs in a Semaphore waiting list then there are Tasks in the system.
3. All Tasks which are waiting on a Semaphore waiting list are of the same type.
4. All Tasks which issue requests to the Kernel Task are valid Tasks, i.e. the Kernel Port Waiting List must only contain TaskIds.
5. All requests on the Kernel Port are sorted in order of their priority.
6. All entries of the Semaphore Waiting Lists must be valid TaskIds.
7. All Semaphore requests are sorted in order of the priority of the issuing Tasks.
8. No Task waiting for a Semaphore can be ready, i.e. it cannot be on the Ready List.
9. No Task on the ready list can be waiting for a Semaphore.

10. Any Task that is ready cannot be on the Semaphore waiting list, i.e. no task ID on the Ready List may at the same time be on any Semaphore waiting list.
11. The Semaphore Waiting List must be empty whenever the semaphore Count is larger than zero.
12. Only Packets of type "SID ReceivePacket" are allowed on the waiting list, i.e. Packets to signal the Semaphore must never be put on the Semaphore Waiting List.

5.3.9 Proof Obligations

The following lists the proof obligations, derived from the properties listed in Sect. 5.3.8, for the Semaphore model:

1. There are never more task IDs on the Ready List than there are Tasks in the system:
$$Cardinality(ReadyList) \leq Cardinality(TaskId) \qquad (5.22)$$

2. There are never more task IDs in a Semaphore Waiting List then there are Tasks in the system:

$$\forall p \in SemaphoreId : Len(SemaphoreWL[p]) \leq Cardinality(TaskId) \quad (5.23)$$

3. All Tasks which are waiting on a Semaphore Waiting List are of the same type:

$$\forall p \in SemaphoreId :$$
$$\forall i, j \in 1..Len(SemaphoreWL[p]) :$$
$$PreallocatedPacket[SemaphoreWL[p][i]].type =$$
$$PreallocatedPacket[SemaphoreWL[p][j]].type \qquad (5.24)$$

4. All Tasks which issue requests to the Kernel Task are valid Tasks, i.e the Kernel Port Waiting List must only contain TaskIds[2]:

$$\forall i \in 1..Len(KernelPortWL) : KernelPortWL[i] \in TaskId \qquad (5.25)$$

5. All requests on the Kernel Port are sorted in order of their priority:

$$\forall i, j \in 1..Len(KernelPortWL) :$$
$$(i \leq k) \implies (PreallocatedPacket[KernelPortWL[i]].prio \leq$$
$$PreallocatedPacket[KernelPortWL[j]].prio) \qquad (5.26)$$

---

[2]How safe is this statement if we use integers to differentiate Tasks and Packets, and other things?

6. All entries of the Semaphore Waiting Lists must be valid TaskIds:

$\forall p \in SemaphoreId :$

$\quad \forall i \in 1..Len(SemaphoreWL[p]) : SemaphoreWL[p][i] \in TaskId$ (5.27)

7. All Semaphore requests are sorted in order of the priority of the issuing Tasks:

$\forall p \in SemaphoreId :$

$\quad \forall i,j \in 1..Len(SemaphoreWL[p]) :$

$\quad\quad (i \leq j) \implies (PreallocatedPacket[SemaphoreWL[p][i]].prio \leq$

$\quad\quad PreallocatedPacket[SemaphoreWL[p][j]].prio)$ (5.28)

8. No Task waiting for a Semaphore can be ready, i.e. it cannot be on the Ready List:

$$\forall p \in SemaphoreId :$$

$$\forall i \in 1..Len(SemaphoreWL[p]) :$$

$$SemaphoreWL[p][i] \notin ReadyList$$ (5.29)

9. No Task on the Ready List can be waiting for a Semaphore:

$$\forall t \in ReadyList : \forall p \in SemaphoreId :$$

$$\forall i \in 1..Len(SemaphoreWL[p]) : SemaphoreWL[p][i] \neq t$$ (5.30)

10. Any Task that is ready cannot be on the Semaphore Waiting List, i.e. no task ID on the ReadyList may at the same time be on any Semaphore Waiting List:

$\forall t \in TaskId$

$\quad (t \in ReadyList) \implies (\forall i \in SemaphoreId : \forall j \in 1..Len(SemaphoreWL[i]) :$

$\quad PreallocatedPacket[SemaphoreWL[i][j]].RequestingTaskID \neq t) \vee$

$\quad (\forall i \in 1..Len(KernelPortWL) :$

$\quad PreallocatedPacket[KernelPortWL[i]].RequestingTaskID \neq t)$

(5.31)

11. The Semaphore Waiting List must be empty whenever the Semaphore Count is larger than zero:

$$\forall e \in SemaphoreId :$$

$$(Count[e] > 0) \implies (Len(SemaphoreWL[e]) = 0)$$ (5.32)

12. Only Packets of type "SID_ReceivePacket" are allowed on the Waiting List, i.e. Packets to signal the Semaphore must never be put on the Semaphore Waiting List:

$\forall e \in SemaphoreId :$

$\quad List\_isEmpty(SemaphoreWL[e]) \implies$

$\qquad (\forall i \in 1..Len(SemaphoreWL[e]) :$

$\qquad PreallocatedPacket[SemaphoreWL[e][i]].type =$ "SID_ReceivePacket"

$$(5.33)$$

5.3.10   Checking the Models

Before checking a model a correct configuration file (*.cfg) is needed. The configuration file allows to check model the model with different parameters and take into account more or less state.

To run TLC for checking the model, enter into model's directory and type:

```
java tlc.TLC name

name - name of the TLA model file

Example:
cd Semaphore
java tlc.TLC Semaphore.tla
```

5.4   Model Verification

Contrary to a typical formal verification approach, in the OpenComRTOS project formal models were developed prior to any implementation. As such these formal models were used to support the architectural design for example to discuss the required properties or the algorithm to be used. The benefit of this approach is that the engineers have a common abstract model to reason about in absence of implementation artefacts. In addition, the models were incremental allowing to start from a simple, very abstract model with few details to one that at the end was very close to what would be implemented. The benefit was that each incremental model was already verified before the next slightly more complex one was developed. This also gave fast turn around times between discussions. Another benefit is that once the models were finalised a first running implementation was reached in very short time, basically because most design decisions had already been taken and improved upon. The resulting code was also very stable and efficient because the formal modelling resulted in a clean architecture, much cleaner than what can be obtained with ad-hoc bottom-up coding.

Nevertheless, implementation is to be done in an available programming language (ANSI C in our case) using an available compiler. There is no a priori guarantee that the implementation is still provable. For this reason a reverse operation was undertaken after implementation: new models were developed that reflected the implementation and model checked again. Very few issues, mostly due to implementation choices, were found and given the clean architecture, model checking of the different entities was straightforward.

5.5 Conclusion

Modelling the OpenComRTOS Entities drives home the point that formal modelling is a very important part of software development. Invariant properties we have checked with the models prove that the system needs only one waiting list in each Entity. Furthermore, we showed that the Semaphore does not have deadlock and matches all properties we required.

But modelling with TLA$^+$ has some drawback. In a system with many constants or with large valued constants, we experience state explosion and the checker can not finish the calculation on a PC or any other processing machine.

In view of the above it is clear that verification of the properties of systems must combine several approaches. The next step of formal modelling of OpenComRTOS Entities is using automated theorem proving like used in abstract interpretation approaches (see, for example, Cousot (2008)). In contrast to TLA$^+$ modelling, this approach is based on symbolic interpretation using axioma's and theorems. To some extent, model checkers are sophisticated simulators that traverse the whole concrete state space and verify that all invariant properties remain valid. In reality, this is not needed especially when numerical values are processed. For the model checker every bit in the numeral value corresponds to a different state. In reality, often only the boundary values can result in a violation of the properties.

Chapter 6
Final Architecture of the RTOS

This chapter discusses the implementation architecture of OpenComRTOS. This is done by using a small test program, called the Semaphore Loop. It follows a top down approach by first explaining the Semaphore Loop itself, including a definition of the Semaphore in terms of OpenComRTOS. This is followed by an explanation of what happens when a Task sends a request to the OpenComRTOS Kernel, taking as example the signalling and testing of a Semaphore. This example then is extended to a system consisting of two processing nodes, connected via a link, each of which executing part of the Semaphore Loop example. While the logical behaviour is the same, the resulting changes in the execution are discussed.

6.1 The Building Blocks of OpenComRTOS

Before explaining how the Semaphore Loop works in OpenComRTOS, it is necessary to first explain the basic elements OpenComRTOS consists of:

1. Tasks – they do the actual work in OpenComRTOS, any computation is performed in the context of a Task.
2. Hubs – a form of a guarded action. Tasks synchronise and communicate using intermediate Hubs. Hubs are available in different types like Events, Semaphores, Resources, FIFOs, Packet and Memory Pools.
3. Packets – while most of the time not directly visible to the application software engineer, Packets are used in OpenComRTOS as the main datastructure at the system level.

An OpenComRTOS program is an implementation model of a more abstract architectural model that is composed of entities and interactions. In OpenComRTOS, we mainly have two types of entities: Tasks and Hubs. Tasks are sequential program segments that can be pre-empted (by a higher priority Task) separated by de-scheduling points. The latter are points in the execution of a Task where

E. Verhulst et al., *Formal Development of a Network-Centric RTOS: Software Engineering for Reliable Embedded Systems*, DOI 10.1007/978-1-4419-9736-4_6,
© Springer Science+Business Media, LLC 2011

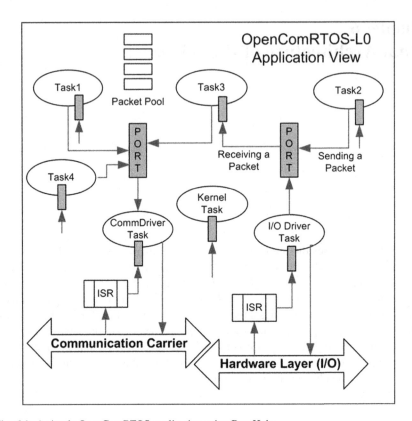

Fig. 6.1 A simple OpenComRTOS application using Port Hubs

they can interact with other Tasks through Hubs. Hubs have specific semantics, mainly consisting of a guarded synchronisation step followed by an action when the synchronisation predicate has been satisfied. Implementation wise, OpenComRTOS uses Packets as containers for the synchronisation and communication data.

Figure 6.1 on page 108 depicts an OpenComRTOS application consisting of a number of Tasks, different Port Hubs and a Packet Pool.

The following sections detail each of these core elements.

6.1.1 The Hub Entity of OpenComRTOS

6.1.1.1 What Is the OpenComRTOS Hub?

Tasks synchronise in a Hub following specific semantics. At an abstract level, a Hub can be seen as a guarded interaction entity. The functional code for this behaviour is executed in the context of the Kernel Task (the highest priority Task

Fig. 6.2 Hub diagram

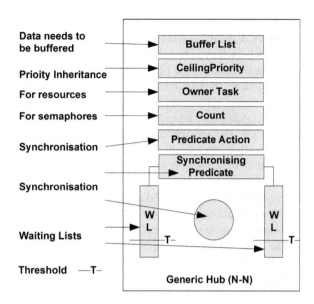

in the system) and hence is executed in a protected section guaranteeing state consistency. OpenComRTOS provides a number of predefined Hubs corresponding to what is commonly provided in most RTOS, like Events, counting Semaphores, FIFOs, Resources, Ports, Memory Pools and Packet Pools. In principle, the user can create his own Hub entities as long as they adhere to the hub semantics. Real-time constraints however dictate that such Hub functionality must be kept short in duration as its execution in the context of the Kernel Task impacts on the scheduling latency of higher priority Tasks.

Figure 6.2 on page 109 is a visual representation of a generic OpenComRTOS Hub.

6.1.1.2   How a Hub Can Be Used

When programming a concurrent system, there is the implicit assumption that the program is composed of independent entities, often called processes or Tasks as in the case of OpenComRTOS. These Tasks also have to interact to achieve a common behaviour at the system level. While they interact, some information is exchanged. What OpenComRTOS achieves is that this happens in a target topology independent way. Therefore, the Hub entities have been decoupled from the interacting Task entities. The behaviour is exactly the same whether the Tasks and Hubs are placed on the same processing node or whether they are placed on different nodes. We call this the logical behaviour. On the other hand, the timing behaviour will differ but this also applies when e.g. the node processor is changed to another type or is running with a different clock rate or memory configuration.

What type of Hub is used depends on the semantics of the behaviour one wants to achieve or needs to achieve. For example, a simple synchronisation on a boolean event can be achieved by using an event Hub. The boolean event semantics however impose restrictions if the behaviour needs to be consistent at all times. Tasks only synchronise when the boolean event is true, hence in all other cases, they must block ("keep waiting"). The semantics also imply that each event is associated with a very specific condition in time. If this condition can be relaxed, a counting semaphore can be used. The semaphore counter will "remember" that the event has happened hence the signalling Task can continue without blocking. On the other hand this implies that all semaphore events are "equal" and no assumption should be made on when the semaphore was signalled.

The Hub entities can provide more complex semantics as well. For example, with a FIFO Hub data will be transferred and buffered until the FIFO buffer is full. Using a Resource Hub, logical critical sections can be created.

The Hub implementation achieves this as follows: When a Task reaches a descheduling point, it issues a Kernel service request, for example it wants to signal a semaphore. While at the application level this looks like a function call, in reality this function call will use a preallocated Packet, composed of a header and a payload section. The header fields that are relevant for the service request are filled in and the Packet is passed on to the Kernel Task. The Kernel Task will become active (as it has a higher priority) and it was waiting on a service request on its Kernel input Port (this is the start of the Kernel loop). The Kernel Task will inspect the Packet header and see that it concerns a particular Hub. The Kernel will inspect the state of the Hub and verify the synchronisation predicate. For example, if another Task is waiting on the semaphore to be signalled, then this other Task becomes active again. If not, the semaphore count is incremented and the signalling Task is made active again. The semaphore counter allows us to introduce a degree of asynchronous behaviour. At least as long as the maximum counter value has not been reached.

The table below gives an overview of the standard Hub types supported in OpenComRTOS. Note that this describes only part of the functionality and only the case with waiting interaction semantics.

6.1.1.3   The Hub as Generic Programming Concept

As a Hub is a generic entity it has a generic structure. It will have some generic elements like a waiting list, additional helper variables for extending the generic behaviour (for example a boolean for event handling, a counter for a semaphore) and two functions. The first is the synchronisation predicate. It is checked whenever a service request is made to the Hub. If the synchronisation predicate is true, the synchronisation action is called. For example, to make the waiting Task active again. A specific type of Hub is a Port Hub. The Port Hub is used to exchange Packets between a sending Task and a receiving Task. When a send request and a

Hub type	Request type	Guard	Action
Port	Put	Waiting Get request	Both Task rescheduled, Packet exchanged
Port	Put	No waiting Get request	Task enters WAIT state
Port	Get	Waiting Put request	Both Tasks rescheduled, Packet exchanged
Port	Get	No waiting Put request	Task enters WAIT state
Event	Put	Event = FALSE	Event = TRUE, Task rescheduled
Event	Put	Event = TRUE	Task enters WAIT state
Event	Get	Event = TRUE	Event = FALSE, Task rescheduled,
Event	Get	Event = FALSE	Task enters WAIT state
Semaphore	Signal	Semaphore count <MAXINT	Semaphore incremented, Task rescheduled
Semaphore	Signal	Semaphore count = MAXINT	Task enters WAIT state
Semaphore	Get	Semaphore count >0	Semaphore decremented, Task rescheduled
Semaphore	Get	Semaphore count = MAXINT	Task enters WAIT state
Resource	Lock	Resource has no owner Task	Task becomes owner, Task rescheduled
Resource	Lock	Resource has owner Task	Task enters WAIT state, priority inheritance applied
Resource	Unlock	Resource has no owner Task	Task rescheduled, return code RC_FAIL
Resource	Unlock	Resource has owner Task	Task rescheduled, return code RC_FAIL if owner Task different from self
FIFO	Enqueue	Count FIF0 entries between 1 and maximum	Task reschedules, data enqueued
FIFO	Enqueue	Count FIF0 entries = maximum	Task enters WAIT state
FIFO	Dequeue	Count FIF0 entries between 1 and maximum	Task reschedules, data dequeued
FIFO	Dequeue	Count FIF0 entries = zero	Task enter WAIT state
Packet Pool	Get	Packet available	Task reschedules, Packet removed from Pool
Packet Pool	Get	No Packet available	Task enters WAIT state
Packet Pool	Put		Task reschedules, Packet returned to Pool
Memory Pool	Get	Memory block available	Task reschedules, block removed from Pool
Memory Pool	Get	Memory block available	Task enters WAIT state
Memory Pool	Put	Memory block available	Task reschedules, block returned to Pool

receive request are both present – we say we have a matching pair – both Tasks are made active again. The Packets are also used to keep track of the waiting Tasks in the waiting list. From above, it is trivial to see that by decoupling the Hubs from the Tasks (and hence also from the Kernel Task even if the Hub functions are executed in the context of the Kernel Task) and by decoupling the Hub datastructure from the Hub functions, one can essentially create any type of Hub, independently of the Kernel Task of OpenComRTOS. It is sufficient to create a Hub with the necessary helper variables, to define the service request and its semantics and the two Hub functions. For example, one could create a Hub that when a sensor Task detects that a certain average temperature has been reached for a certain period of time (for example by keeping track of the last 3 time-stamped measurements), a function is called that activates a motor controller Task with specific parameters. What we witness here is that the OpenComRTOS Hubs allows to develop an application specific concurrent programming environment. This was made possible because the formal development gave insight in the architecture and functionality so that the generic mechanisms could be discovered. As a result, even the Kernel Task does not need to be recompiled when adding a user defined type of Hub.

6.1.1.4   The Special Case of the Port Hub

In a pure CSP context, processes synchronise using blocking channels. Upon synchronisation, information can be passed over the channel. The Hub generalises this concept at several levels:

- Tasks do not need to be blocking, allowing also for non blocking, blocking with time-out and asynchronous semantics.
- Hubs are decoupled from the Tasks, allowing arbitrary Tasks to synchronise over a given Hub (we call this N to N semantics).
- Hubs allow to synchronise on any synchronisation predicate as long as it is a valid boolean expression.
- The resulting action can be any function valid in the application and OpenComRTOS domain.
- Hub interaction is independent of the topology of the underlying node network and mapping. This is also true for pure CSP as it is an abstract process algebra, but often not for practical implementations like occam.

The Hub type in OpenComRTOS that comes closest to a CSP channel is called a Port. Tasks can send a Packet to a Port and Tasks can receive a Packet from a Port. Whenever there is a matching send–receive pair, synchronisation happens and the Packet is passed from sender to receiver Task. Note that in the implementation, only the relevant fields (like the data) are copied while the Packets remain linked with the original requesting Tasks. Hence, in principle such a Port Hub can be used to emulate almost any other of the Hub interaction types. If no data is passed, it acts like an event or semaphore depending on whether synchronous or asynchronous semantics are used. If data is passed as well, it can act like a FIFO Hub. It can even be used as simple Resource lock by passing a Packet around as a token.

So why create other Hub types as well? The main reason is that when using the Port Hub, the application developer must add himself the additional semantics and remember what he wanted to achieve as behaviour. For example, he can add a type field but that type field has to be tested upon in the receiver Task. Hence, for standard Hub types, the little extra code in the implementation outweights the drawbacks. The code becomes more readable and it is easier to generate trusted code automatically.

6.1.2   Tasks

A Task in OpenComRTOS can be compared to a process or thread in common operating systems. This statement is not fully correct and is only made to give the reader a rough idea what a Task is. In truth a Task in OpenComRTOS is not a thread, it is much closer to a CSP Process. In OpenComRTOS, all the application processing is done in the context of a Task, this means that the Kernel, the drivers and the user applications are all Tasks. This is comparable with a microkernel architecture, but given the very small size of the Tasks in OpenComRTOS, one could call it a nanokernel architecture.

6.1.2.1   The Kernel Task

The Kernel Task is, like most other Tasks, implemented as a never terminating Task. It waits until an application sends a request (using a Packet) to the Kernel input Port. The Kernel Task retrieves the Packet from the Kernel input port. If the Packet is addressed to a local entity the Kernel Task will inspect the Packet and call the corresponding services, for example to inspect the Hub state and call the corresponding Hub functions as outlined above for the event and semaphore Hub. If the entity is on another node in the network, the Kernel router is called to determine over which link to send the Packet to bring it closer to its destination and then forwards it to the corresponding link-driver Task.

Although it is possible to develop the RTOS with the Kernel Task not having the highest priority (e.g. assigning the highest priority to a driver Task), in practice it simplifies the implementation. In addition, any Task interaction requires the use of the Kernel Task and assigning it the highest priority will reduce overall latency.

6.1.2.2   Link-Driver Tasks

A link-driver Task controls a point-to-point communication channel, over which it exchanges Packets with another node which runs OpenComRTOS. This forms the basis for the Virtual Single Processor Programming model of OpenComRTOS.

Such a driver Task is simple as its only function is to receive and to send Packets over a physical communication medium. Of course, depending on the hardware this can be rather complex especially if the hardware (often a shared medium in such cases) handles several simultaneous Packet transfers. In that case the point-to-point communication is virtual. The benefit is that OpenComRTOS can make use of almost any type of communication medium, including virtual ones, for example by tunnelling through existing network infrastructures. This is the case for ethernet based TCP/IP based communication. Also when shared memory is the communication medium, OpenComRTOS will use it as set of virtual communication links, avoiding many of the issues shared memory presents to the developer.

6.1.2.3 Application Tasks

An application Task, sometimes called a user Task, will be specific for a particular application and hence is often developed for every application specifically, although the OpenComRTOS development environment allows to define service modules that can be reused across multiple applications. Application Tasks are, hence, the weak links in terms of reliability as they are most of the time not formally developed. Application Tasks are also the ones that need to be composed and scheduled so that together they achieve the specified behaviour. The sequential code segments can be verified using standard formal verification tools whereas scheduleability analysis requires specialised tools that take their input from profiling the application.

6.1.2.4 Idle Task

The idle Task is the lowest priority Task on each processing node. It is automatically scheduled when none of the other Tasks are all de-scheduled. The system then looks idling, hence the name. This Task still consumes CPU cycles, but it can be modified to perform power management Tasks like putting the processor in one of its low power modes or it can be used with a calibrated loop to indirectly measure the processor load.

6.1.3 Packets

OpenComRTOS is designed and implemented as a Packet switching system. Packets are not only used to carry data and information, they are also used to keep track of the system status and to transfer control from one module to another. Packets are used consistently at all levels. A Packet in OpenComRTOS consists of a header- and a data-part. The header-part contains the addresses of the Task that issues the service request and of the Hub the Task will use to interact with another Task of with the rest of the system. Each Task in the system has one Packet available which it can use to issue service requests, hence it is also referred to as a RequestPacket.

A Task sends each request to the Kernel input Port, after which it is de-scheduled and the Kernel Task becomes active to process the request. The Kernel Task then processes the Packet and forwards it to the Hub the Packet is destined to go to. The Hub functions will be called in the context of the Kernel Task after which the Packet is returned (with a return code) to the Task that issued the service request.

In case the Hub is located on a different node than the Task that wants to request a service from the Hub, the Kernel passes the RequestPacket to the corresponding linkDriver, which translates the Packet into a standardised TransferPacket and sends it over its link to the destination node. Once at the destination node the TransferPacket gets translated into a local Packet and injected into the local Kernel Task. The request is then handled like a local request described above.

The Packet organisation is standardised across all nodes within an OpenCom-RTOS Virtual Single Processor system. This allows the developer to construct heterogeneous systems where different types of processors cooperatively work on one problem.

6.2 The Semaphore Loop

Figure 6.3 on page 115 shows two user Tasks (T_1 T_2), synchronising using two semaphores (S_1, S_2).

Listings 6.1 and 6.2 show the source code for the two Tasks that represent the semaphore loop in the application Diagram of Fig. 6.3.

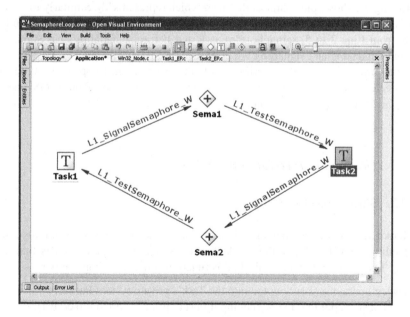

Fig. 6.3 Application diagram with all Interactions for the Semaphore Loop

```
  void T1 (L1_TaskArguments Arguments){
2     while(1) {
          L1_SignalSemaphore_W(S1);
4         L1_TestSemaphore_W(S2);
      }
6  }
```

Listing 6.1 Souce code for Task T1

```
  void T2 (L1_TaskArguments Arguments){
2    while(1) {
         L1_TestSemaphore_W(S1);
4        L1_SignalSemaphore_W(S2);
     }
6  }
```

Listing 6.2 Source code for Task T2

Listing 6.1 shows the code for Task T1 which represents T_1. The Arguments of the function call are not used. All the work is done within the infinite loop, starting from Line 2. The code block signals semaphore S1 before it tests semaphore S1. Both signalling and testing of the semaphores are done using waiting semantics, indicated by the postfix '_W'.

Listing 6.2 shows the source code for T2 which represents T_2. Similarly to T1 the Task is represented by a function whose parameters are not used. Within the while-loop, from Line 2 to 5, semaphore S1 is tested before semaphore S2 is signaled. Again both testing and signalling are done using waiting semantics.

After this high level description of the Semaphore-Loop it is time to investigate what happens during a system call in OpenComRTOS using L1_SignalSemaphore_() as example.

6.2.1   The Semaphore Loop in Detail

6.2.1.1   Single Processor Case

A Kernel service request like signalling or testing a semaphore is essentially a de-scheduling point for the calling Task. What essentially happens then is that the calling Task issues a function call. In this function call three things happen:

- The call parameters are copied into the Task's private system Packet.
- The Packet is put into the Kernel input Port (a priority sorted linked list).
- Context is switched to the Kernel Task.

This mechanism is used consistently when Tasks interact, even when the service request comes from an Interrupt Service Routine (ISR) that has no context on itself. In the latter case the Packet is inserted without waiting, imposing the restriction that only non-waiting services are meaningfully used.

As a result the Kernel Task becomes active and will process all Packets on its Kernel input Port in order of priority. Note here that we are in a multiprocessor context and that Packets coming from other processing nodes might have been added by the driver Tasks. In this case the Kernel will inspect the header fields of the Packet and determine the Hub and its location in the network. If local, it will verify the synchronisation function of the Hub and then call the synchronisation action. As a result, the semaphore count will be incremented. If there was a waiting Task, the semaphore count will decremented. Depending on the outcome, header fields are modified (e.g. with a return code), the Packet is inserted in the requesting Task's input Port and the Task is made active again. If it is still the highest priority one on the ready list, the Kernel Task will switch context to it before de-scheduling itself. When the requesting resumes execution, the service request will return with the return code.

6.2.1.2 Multiprocessor Case

What happens when the service requests involves Tasks and Hubs on remote processing nodes? This is actually quite simple. The destination node field (that is part of the entity's identifier) is used to look up how the node can be reached. Often, this simple means that a communication link is returned. The Packet is then inserted in the Task input Port of the driver Task handling this link and when this driver Task becomes active it simply copies the Packet into a transfer Packet and puts it onto the communication link. At the receiving side, the contents will be received into another local transfer Packet that is then passed on to the Kernel Task input Port and we are back in the single processor scenario.

6.2.1.3 Why Priority Sorting and the Use of Packets Is Important

In a real-time system, priority is used to determine when a Task will run. In combination with scheduleability analysis and real-time profiling, this provides for an effective means to meet hard real-time constraints. There are however a few real world issues that are not very well covered in the theoretical models. For example, in single processor systems the use of cache memories will result in time distribution for the execution time of Tasks. Practically speaking, this means that often the application becomes soft real-time unless one designs only with the Worst Case Execution Times (WCET). Given the growing speed gap between CPU's en memory (a factor of over 100 on GHz range processors), one can see that for real-time applications the cache is often better disabled and replaced with local memory (typically SRAM) that has known (typically fast) timings at all times.

In a multiprocessor system, other phenomena will influence the timing behaviour. Communication media are shared between processing nodes and while in use, they block other Tasks introducing scheduling delays. Therefore prioritised Packets provide relief. First of all as Packets have a finite length, the blocking time can be limited by adjusting the size of the Packet. Smaller Packets will however result in lower Task-to-Task bandwidths. If in addition, the Packets are transmitted in order of priority we guarantee that the highest priority Task are never blocked longer than the time a Packet is transmitted.

Prioritisation is also needed for all waiting lists and Task input Ports (specifically the Kernel Task) because in a multiprocessor system, service requests can come from any node. Hence, multiple requests can be present simultaneously. By sorting them in order of priority, we assure that the highest priority Tasks will always experience least latencies. A particular requirement for any real-time system is support for priority inheritance. This will boost temporarily the priority of a Task that is blocking a high priority Task. While this is trivial to implement in a single processor RTOS, in a distributed RTOS this will affect the priority of Tasks on other nodes and requires as well that waiting list are resorted. This is also supported in OpenComRTOS.

Note that a careful analysis is needed to determine at design time the expected timings. This is outside the scope of this project. Nevertheless, it is clear that good profiling can remove many of the uncertainties. It should also be clear that shared communication media with complex arbitration policies should be avoided as well. A simple point to point communication is often inherently faster and simpler, hence using less energy as well.

6.2.2   Heterogeneous Multiprocessor Systems and Their Issues

In the architecture of OpenComRTOS, no assumption is made about the target system. In principle a system can be composed of several types of processors that are physically connected over several types of communication media. When implementing it on real target systems, some dependencies have to be taken into account:

First of all, given the wide availability of C compilers on embedded targets, a stable and reliable ANSI-C compiler must be available. Nothing prevents implementing OpenComRTOS using other programming languages, at least as long as they have themselves no hidden default runtime system.

Data representation differences, while the C datastructures use standard C types, C compilers will handle them slightly differently from processor to processor. There is the famous little endian versus big endian issue, but there are also data alignment issues. All such differences must be taken into account to guarantee the portability across different compilers and processors but also to guarantee that when Packets travel across a network of processors, they are interpreted correctly on all.

An important issue is the communication infrastructure. In order to improve the real-time behaviour in a distributed system, priorities act system-wide. Hence, while the communication is Packet based, each Packet carries a priority inherited from its generating Task. As a result, mainly the communication delay of one Packet will put a lower boundary on the scheduling latency introduced by interprocessor communication. A major issue however is that some communication media that can be shared by multiple processing nodes are blocking or do not allow priority based arbitration. This is the case for the well known ethernet based communication that in addition suffers heavily when a certain communication load is surpassed. Note however that most of these problems disappear when the communication is reduced to point-to-point connections. The drivers can then perfectly put the Packets in order of priority, minimising any access conflicts.

6.3   OpenComRTOS Development Process for Applications

Given that OpenComRTOS separates the logical behaviour from the temporal behaviour, in principle any target processor can be used to develop an application. For OpenComRTOS two target systems are of particular interest as they can be used for development, simulation as well as real nodes in a system. These systems are a port of the RTOS on top of MS Windows (Win32) and Posix (Linux). A typical development cycle will hence start by developing the program on a single Win32 or Posix node. Additional vertical nodes can be added and Task and Hubs can be remapped to them. The benefit of developing in such an environment is that it is well supported by development tools and no harm can be done when an error is made. Real data input and output can also be simulated using wrapper Tasks. Subsequently one or more embedded node can be attached to the Win32 or Posix machine and Tasks and Hubs can be remapped to them. Another benefit of this approach is that such a Win32 or Posix node can be kept in the system providing "host" services to the often memory constrained embedded target processors. This allows embedded nodes for example to read and write files on the host system, to interact with a human user or to display data graphically. Another benefit is that an OpenComRTOS node can be accessed across the worldwide internet with no changes to the application code.

6.4   Summary

In this chapter, we described the implementation architecture of OpenComRTOS. In particular, we highlighted the use of Packet switching, the generic Hub entity and the specific Port Hub entity. Differences between single processor and multiprocessor functionality were discussed and some important issues related to a multiprocessor implementation were discussed.

Chapter 7
Task Interaction Models in OpenComRTOS

Tasks in OpenComRTOS interact using an intermediate Hub entity. This way the interaction is decoupled from the Tasks themselves. Different temporal semantics (waiting, non-waiting, and waiting with timeout) of Tasks synchronization are formalised. Conclusions are drawn on correct usage confirming the correctness of the original CSP semantics.

7.1 Introduction

Previous chapters discussed the rationale for developing OpenComRTOS and how its architecture was developed using formal modelling. In OpenComRTOS, Tasks represent processing entities and Hubs represent interaction entities. Tasks can only interact through Hubs, while the Hub entities implement the interactions.

The benefit of this approach is that the Hubs decouple the individual Tasks. While a Task is an active entity, a Hub is an entity with a predefined behaviour that mediates between Tasks. By decoupling we mean that a Task does not need to know about the other Tasks it interacts with. During an interaction a copy of part of the internal state of a Task is passed or received and this protects the state of the Tasks. The mechanism is also independent of the location of the Tasks and of the Hubs in the networked or shared memory multiprocessor system, resulting in transparent parallel programming. Although Hubs and Tasks reside at specific memory locations, OpenComRTOS addresses them using their identifying IDs and not using their address pointers. Similarly, when an entity passes data in memory, a copy of its values are passed, not the address pointers. Also data is passed using a copy, although the programmer can if he protects the data accordingly. When optimising, the program can be adapted to exploit the local property that memory is shared and hence OpenComRTOS will not prevent software engineers from passing pointers. Protection can be achieved using resource locking. It is the responsibility of the developer to ensure their validity. They also must be aware that such optimised code can no longer safely be scaled to run on multiple processing nodes.

E. Verhulst et al., *Formal Development of a Network-Centric RTOS: Software Engineering for Reliable Embedded Systems*, DOI 10.1007/978-1-4419-9736-4_7, © Springer Science+Business Media, LLC 2011

All this results in turning an OpenComRTOS Task into a component. By combining components it is possible to create more complex systems glued together by the Hub-interactions. In other words, the interacting entities modelling paradigm gives us a compositional process view. To use a component, we do not need to have access to its internal state – it is sufficient to know its interfaces (i.e. the protocols it obeys).

This view was first formalized in Hoare's Communicating Sequential Processes (CSP) process algebra (Hoare 1985b). In CSP, a system is composed of Processes and Channels. A process executes a possibly infinite number of sequential steps. The sequences of individual steps (called traces of processes) are separated by channel communications. A channel communication can be seen as the simplest form of interaction. It is fully synchronous and when processes synchronize on a channel, data can be transferred over it. This mechanism is very powerful and provides a mechanism for formal construction and verification of large systems. However, the CSP channel communication is simple in its semantics. CSP channels are also time-agnostic although later on this was remedied with the formulation of Timed CSP (Davies and Schneider 1989; Dong et al. 2006).

The Hubs generalize the functionality of a CSP channel. The basic functionality of a Hub is synchronization between Tasks, just like in CSP. A Hub synchronizes Tasks using a boolean guard that can be a lot more complex. The resulting behaviour in CSP is to allow each process to continue. This basic behaviour is also present in the Hub concept. The difference is that a Hub can be specialized because its behaviour is allowed to be user defined. Specific Hub types (i.e. Event, Semaphore, FIFO, Resource etc.) define a well defined superset of the basic CSP semantics. This also allows a Hub to act as an intermediate entity for a larger number of Tasks whereas a CSP channel is a point-to-point connection between two processes only. Note that in CSP the behaviour of a Hub can be achieved as well, by inserting a dedicated intermediate process between Tasks. As discussed, while adequate for formal modelling, from a programmer's point of view abstraction is lost as the mechanism or interaction becomes visible again at the application level.

An important advantage of the Hub is that the interaction behaviour can be customized. This allows the application designer to express the system in a way which matches its intended behaviour. For example, the OpenComRTOS kernel provides standard support for blocking, non-blocking, blocking with a timeout or fully asynchronous interaction. Application specific Hubs can be created by customizing the synchronization and action functions, and this is possible without having to modify the kernel itself. It is also supported by the OpenComRTOS metamodel, used for code generation and for the visual development of applications. This is different from traditional approaches, that either require a complete rebuilding of the kernel, or the creation of a middle-ware layer that emulates the required behaviour using standardized kernel services.

Using guards before actions makes it amendable to formal reasoning, e.g. by using TLA (Temporal Logic of Actions) (Lamport 1983). The Hub structure consists of a logical proposition (the synchronization condition) and the synchronization action. The synchronization action is invoked once the synchronization condition is true. We expand the Hub model by decomposing the synchronization proposition into the pre and the post conditions. In this chapter, we also will show the analogy between Hoare triples $\{P\}\,C\,\{Q\}$ and the Hub.

Section 7.2 of this chapter expands the semantics of the CSP interprocess communication by creating a model of Task interactions via intermediate Hub entities. A sequence of such interactions creates a protocol for intertask interaction. Section 7.3 explores the time properties of the proposed Task interaction model. This is followed in Sect. 7.4 with a detailed description of asynchronous interactions. The conclusions and references sections finalize the chapter.

7.2 Modelling Task Interaction

Dividing a system into entities and interactions is not new. It represents the natural way of human thinking. Accordingly, existing modelling techniques emphasise the use of objects and their relationships. Currently, the most prominent examples are Object Oriented Design (OOD) and Object Oriented Programming (OOP) (Booch et al. 2007). The common object model, that encapsulates the internal properties and methods, however, pays less attention to the interactions between such objects and as result its view is fairly static.

Embracing interacting entities as an architectural modelling approach is much better at expressing the dynamic interactions between objects. By using the term *interaction* we express that *Tasks take actions in a mutual way*, which is often considered just a side effect of their communication. But the mutual way of actions does not mean that Tasks perform actions in cooperation toward some predefined goal. This is the difference to the agent-based approach.

Interactions take place when two or more entities have an effect upon one another. So, the basic idea of interaction consists of a two-way effect, as opposed to a one-way causal effect. Figure 7.1 presents such an interaction scheme.

From this simple model follows that an interaction should incorporate at least two mutually linked actions in opposite directions, named here $Action_1$ and $Action_2$. A Hub also supports N–N Task interactions scheme.

In this model, an interaction between Tasks is composed of two sub-interactions between each Task and the intermediate Hub entity. The sub-interaction between a Task and a Hub also consists of two complementary interactions, that is also reflected in the layered implementation of OpenComRTOS where Packets are used to implement the interactions. An interaction is composed of a send–receive pair of Packets. Each of these send–receive actions can be seen as a request to synchronise via a Hub followed by an acknowledgement, expressing success or failure of the interaction (called return values when using function calling). This is an essential property in the context of embedded real-time systems that must be predictable. Note, that in other approaches the return values of services are not always considered or it is assumed, that the interaction is always successful. Hence, many programmers do not test the return values.

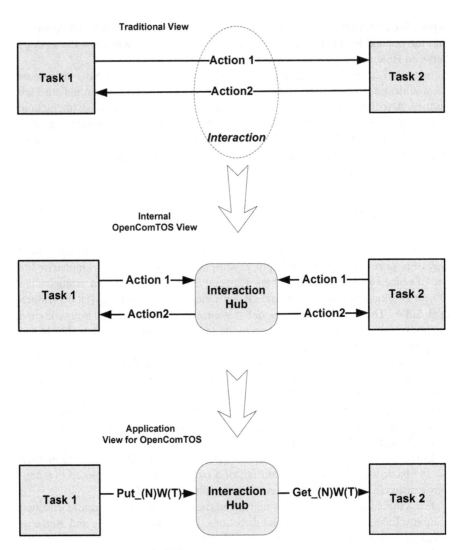

Fig. 7.1 Task interactions with a Hub

To simplify the analysis, we abstract away these sub-interactions and only consider the top level Task interactions. Such a Task interaction consists of two symmetric actions, which we call Put- and Get-actions. These terms are just place holders for the actual Packets, that are interchanged in the Hub. No interaction order is implicitly defined in the names. An interaction is hence divided into a matching pair of Put (that we will call P) and Get (that we call G) actions leading to an effect S, i.e. the synchronization between Tasks. Note that we do not state anything regarding action timing and order, nor are any guards attached to these actions.

We only express that the mutual actions between two Tasks are mandatory for synchronization.

$$(P \wedge G) \vee (G \wedge P) \implies S \tag{7.1}$$

where $\wedge$ represents the logical AND symbolising that both actions must have happened for the synchronization S to take place. We use here the concept of action correspondingly to the TLA definition of (Lamport 2002a, p.16): *"an action is an ordinary mathematical formula, except that it contains primed as well as unprimed variables... An action is true or false on a step."*

Equation (7.1) follows from one of the design principles of OpenComRTOS – the symmetry of the synchronization mechanism. The symmetry principle results in using a synchronisation entity in between Tasks. This principle is used to verify the properties of OpenComRTOS applications – i.e. for a Task synchronization to happen (7.1) must be satisfied.

Tasks in a concurrent system have equal communication rights in the system. Thus by default, there is no master Task, which causes effects upon other Tasks. A Task sending a Packet to a Hub knows nothing about the other Task(s) interacting with this Hub, i.e. *there is no cause-effect relationship between the interacting Tasks.*

The model, defined in (7.1), states that *only the mutual actions of two Tasks lead to its synchronization*, and that's why we classify it as an *interaction*. An important case to consider occurs when only one action (Put or Get) happens. If an external observer detects only the first phase of the interaction, then the synchronization has started, but has not yet finished. In case of blocking services, the Task which initiated the action, is prevented from making progress. In other words, the Task has to wait for the second phase of the interaction. Sometimes this is also called blocking.

Equations (7.2) and (7.3) model the scenario when $Task_1$ executes the blocking action P_1, but $Task_2$ did not yet initiate action G_2. This results in the $Task_1$ being blocked. Similarly, $Task_2$ is blocked after it has executed G_2 and $Task_1$ did not initiate P_1 yet.

$$P_1 \wedge \neg G_2 \implies Task_1 \ blocked \tag{7.2}$$

$$\neg P_1 \wedge G_2 \implies Task_2 \ blocked \tag{7.3}$$

Note that (7.2) and (7.3) reflect the usual semantics of processes communication on a blocking CSP channel[1]. A blocked Task becomes unblocked if the corresponding action occurs. Corresponding means here that the actions types should be matching: if the first action is of the Put type, then it follows that the second action must be of the Get type. Synchronization is the only way to allow blocked Tasks to make progress again.

---

[1] Note, in OpenComRTOS Tasks are never blocked when using non-waiting services.

7.3   Timing Properties of Task Interactions

Equation (7.3) states that only matching actions of Tasks lead to synchronization. In this section, we discuss the meaning of the term matching with respect to time. OpenComRTOS implements three possible time related semantics for interactions using a Hub: Waiting (W), Non Waiting (NW), and Waiting with Timeout (WT). Let's define the effects (i.e. synchronisation or the absence of it) when the sub-actions have different timing properties. In general, the semantics of Task interactions depend on the Hub type and its boundary conditions (e.g. a FIFO list size). To simplify the discussion this section will use the concept of synchronisation in the CSP sense whereby upon synchronisation data is transferred between two Tasks. The OpenComRTOS entity closest to a CSP channel is the Port Hub. Its semantics are that Packets are exchanged between Tasks by sending and receiving Packets. In the following the Port Hub is taken as the example because it is the most general case.

Table 7.1 shows the nine possible cases of time semantics for the interaction of two Tasks.

The W semantics synchronisation is not dependent on time. The symmetry cases, located at the diagonal of Table 7.1, occur when both Tasks use the same type of service. In these cases, synchronisation does not depend on the relative order of the interactions. For completeness, however, we must mention that synchronisation will still depend on the priority, at least when multiple Tasks end up in de waiting lists as a Task is inserted in order of priority in the waiting list. However, this has no impact on the specified logical behaviour.

In case of any of the following enumerated interaction pairs, the synchronization depends on the time ordering of the interactions. Temporal logic operators should be used to express the time semantics of the synchronization mechanism in such a case:

1. <W, WT> – If the W action happens before the WT action, then the synchronization will occur. If the WT action happens first, then to have synchronization the W action must happen inside the time interval T.
2. <W, NW> – If the W action happens before the NW action, the synchronization will occur. If the NW action happens first, then the synchronization will not occur.
3. <WT, NW> – If the WT action happens before the NW action, then to have synchronization the NW interaction should happen inside the time interval T. If the NW action happens before the WT action, the synchronization will not occur.

Table 7.1 Time semantics of two Tasks interacting in OpenComRTOS

Time semantics	W	WT	NW
W	Symmetry	<W, WT>	<W, NW>
WT	<WT, W>	Symmetry	<WT, NW>
NW	<NW, W>	<NW, WT>	Symmetry

By extension, a W action is a WT action with $T = \infty$ and a NW action is a WT action with $T = 0$. Hence, it should be clear that a Task using the NW action in an application is not deterministic as it depends on whether the other Task it synchronises with was waiting or not. Hence, the NW action should not be used unless the application takes into account this uncertainty. For the same reason, an interaction pair <NW, NW> can never synchronise because processors operate in sequence on their instructions.

Note that in any implementation there is a small uncertainty on the interval T. As the actions take time (in the form of a finite number of instructions), the synchronisation will depend on when the T timer event is seen by the kernel Task. For the sake of the analysis, we ignore this uncertainty as it does not impact on the logical behaviour and in a correctly parametrised real-time system this uncertainty will be significantly less than T.

Such basic sequences of interactions can be applied to analyse more complex interactions by means of composition. Therefore, we formulate them as temporal formulas in TLA. The next state of the system will depend on the resulting state of previous interactions (leading to synchronization or to blocking of the application).

Let's formalize and consider the time properties of the symmetry cases of OpenComRTOS synchronisation semantics.

An NW Task interaction has the following semantics:

$$(P_{t=t_1} \wedge G_{t=t_2}) \wedge (t_1 = t_2) \implies S \tag{7.4}$$

Equation (7.4) defines that, in case of NW services, we have synchronization S if and only if both Tasks exhibit their respective actions, P and G, at the same time[2]

Therefore, <NW,NW> synchronisation works under no circumstances - NW services will always fail because of the sequential execution on the CPU (which is a sequential von Neuman machine). The kernel, where the Hub is located, strictly serialises the access to the Hub, therefore, the Hub will never see two complementary NW-requests at the same time, thus NW synchronisation at the Hub is not possible.

Equation (7.4) can only be true when one of actions (of the Put or Get type) was buffered in the waiting list of a Hub. For example, we can get a Packet from a FIFO using NW semantics if it was already put in the FIFO using NW semantics. So the NW synchronisation semantics will only succeed when the interactions are "buffered" in the Hub which transforms the semantics in a waiting one. It expresses the fact that Task synchronisation is a secondary effect from successful completion of the sub-interactions between a Hub and a Task and not directly between Tasks.

Hence, the only safe semantics for interactions is the waiting variant. Therefore, this is the usual way for using Task synchronization in an application program (as e.g. in CSP). The NW services can only be used as a way to check *whether or not synchronization is possible*. This can only be successful when the Hub has a

---

[2]The discrete time t, defines the timestamps of actions P and G.

buffering capacity. Needless to say that using the NW action in a (polling) loop until success is not only wasting processing cycles, but also makes the Task unpredictable in the real-time sense.

WT Task interaction (synchronization) has the following semantics:

$$(P_{t=t_1} \wedge G_{t=t_2}) \wedge (t_2 - t_1 < T) \implies S \qquad (7.5)$$

where T stands for the timeout interval. The interaction for NW and W can be considered as the partial case of WT with $T = 0$ and $T = \infty$.

Equation (7.5) defines that, with WT services the synchronization occurs when the time difference between actions P and G is less than the interval T. Such formula can be applied to NW services, with condition that $T = 0$[3].

This leads to the following synchronization semantics for W Task interaction:

$$(P_{t=t_1} \wedge G_{t=t_2}) \wedge (t_2 - t_1 < \infty) \implies S \qquad (7.6)$$

For the case of using W services, the time, during which Task synchronization can happen, is infinite. So, (7.6) can be simplified to (7.1), i.e. there is no real time dependency – the synchronization occurs in case of matching actions irrespectively of their time. This type of synchronization here is the same as for a CSP channel. Note, however, that W action can also wait forever if the synchronising Task never reaches its synchronisation point. So, for safety programming, one must always use WT actions and test on success or failure.

7.4   Notes on Asynchronous Interactions

Fully asynchronous interactions are also possible in OpenComRTOS, although such interactions are really synchronous interactions whereby the synchronisation is delayed and decoupled from the processing a Task is executing when the interaction was initiated. The semantics of asynchronous-in-time interactions can be defined as two phase interaction semantics in Fig. 7.1. While in the previous section the Task is made waiting between the two sub-actions (hence it is a one-phase interaction), when using an asynchronous interaction, the Task can continue after it has issued the first sub-action to synchronise later on. The latter is mandatory because real systems always have limited resources whereas asynchronous interactions could require an unlimited number of resources (e.g. for data storage). This imposes serious limitations on the verifiability of applications. Therefore, asynchronous interactions must be restricted and must resynchronise before the resources are depleted. Figure 7.2 shows how asynchronous interactions between Tasks and Hubs take place. Asynchronous interactions were not fully formally verified in this project

---

[3]In real systems the communication delay defines a minimum meaningful timeout value.

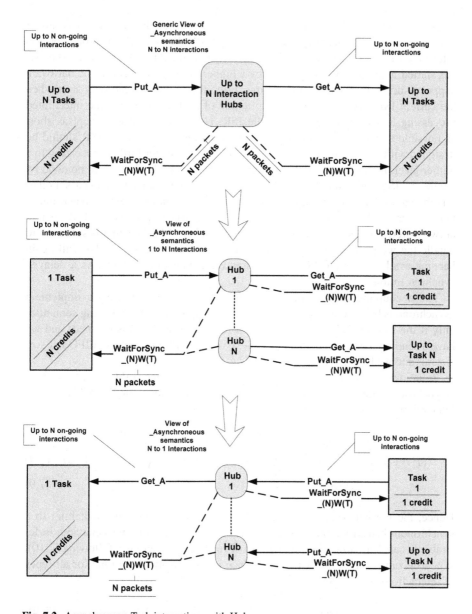

Fig. 7.2 Asynchronous Task interactions with Hubs

and hence an in-depth discussion is left out. Nevertheless, it is useful to discuss them further as it highlights some issues mainly in the context of interacting with the processing hardware.

On a physical processor, the software level has to interact with the real-world environment. Often this will be achieved by using interrupt signals at the hardware level. In general, the hardware implementation will force the processor to transfer control to a specific Interrupt Service Routine (ISR). We say that the processor was interrupted and this happens at a higher priority than the RTOS kernel. Once the ISR has started it can only be interrupted by a higher priority interrupt (if the hardware supports such prioritisation of interrupts, else other interrupts can only start when the current ISR has terminated). Termination of an ISR to resume the interrupted processing is also under control of the ISR. Hence, ISRs must be kept as short as possible as they will block any other processing. Therefore, an ISR should be designed with the hardware restrictions in mind and transfer further processing to a higher level, but pre-emptive driver Task. In OpenComRTOS, this can be done by having the ISR collecting the interrupt relevant data in the a Packet and pass it on to a Hub, e.g. an Event or Port Hub. An ISR, however, cannot 'wait' to synchronise, hence it inserts the Packet asynchronously. As the driver Task and the ISR operate at two different rates, the ISR needs to use a number of Packets that is sufficient to handle the worst case. When the driver Task synchronises, the Packet will again become available for the ISR and we can effectively consider this as a delayed synchronisation.

The decoupling functionality of a Hub provides another way to implement asynchronous behaviour. For example, for a FIFO, we will have waiting semantics only if the FIFO is full (the limit of the FIFO list size ($Size$) is reached) and we have a sending Task: ($P \wedge Count = Size$)); or if the FIFO is empty (FIFO list size is zero) and we have a receiving Task: ($G \wedge Count = 0$).

Equations (7.7)–(7.9) detail the conditions when the FIFO converts from Asynchronous Interactions back to synchronous behaviour, for W, WT, and NW interactions.

$$(P \wedge Count = Size) \vee (G \wedge Count = 0) \implies Wait \qquad (7.7)$$

$$((P \wedge Count = Size) \vee (G \wedge Count = 0)) \wedge (|t_2 - t_1 < T|) \implies Wait \quad (7.8)$$

$$(P \wedge Count = Size) \vee (G \wedge Count = 0) \implies \neg S \qquad (7.9)$$

Hence, the normal behaviour of such a service is asynchronous, switching to a synchronous behaviour only when the FIFO is full (or empty). This behaviour is desirable because it allows limiting the use of resources. Otherwise, following (7.10), synchronization on a FIFO Hub takes place irrespectively of the time for all types of interactions (i.e. W, NW, WT).

$$\neg((PwedgeCount = Size) \vee (G \wedge Count = 0)) \implies S \qquad (7.10)$$

Outside this asynchronous domain, we have all restrictions of the synchronous Hub services: NW services are not to be used and waiting is in order of priority.

Similar properties can be formulated for the Event[4] and Semaphore Hub functionality, which also exhibits an asynchronous behaviour. For the Event Hub, $size = 1$ and for the Semaphore, $size$ is equal to an implementation specific large number (typically $2^{32} - 1$ for a 32bit processor). But as these Hub interactions do not pass data, the buffer space can be reduced to the counter value. The difference is that for an Event the maximum $Count = 1$ (representing a Boolean) and for a Semaphore Hub the maximum $Count$ is machine dependent.

In the same context, it was also decided not to implement a Hub with the functionality of a "pipe" (as found e.g. in Unix systems) because the pipe semantics assume an infinite buffer to be available, resulting in a system failure, if the receiving side becomes blocked.

7.5 Conclusions

This chapter discussed OpenComRTOS as an interaction oriented programming paradigm. The proposed approach results in the unification of different forms of synchronization methods into the unique Hub concept. The Hub allows the application designer to express the synchronization mechanism as best matching the behaviour of a system, for example, by using different forms of time synchronization semantics or by defining domain specific synchronisation predicates and actions. Building the Hub on a formalised base links programming techniques with formal methods, and therefore, offers the possibility of designing formally proven programs.

---

[4]The Event Hub is a Semaphore Hub with a maximum count value of 1.

Chapter 8
Results: Code Size and Performance

We briefly show that the resulting architecture generates very small code (5 KB for the distributed version) resulting in better performance and better safety properties.

8.1 Metrics of Success

When developing an RTOS or porting it to a new processor, it is paramount that the RTOS is "stable". In other words, when the application is correctly developed, it can in principle only fail because either the environment injected erroneous data (but a safe application should anticipate this) or the kernel still has an unknown error. The purpose of the formal development of OpenComRTOS was to reduce this probability to a minimum. On real targets however, there are still other dependencies. For example, the compiler can generate erroneous code. The latter can happen in particular when optimisation switches are enabled. Some of the errors will be due to data alignment issues but with careful coding the kernel implementer can force the compiler to always generate correctly aligned data structures. A final source of failures has its root cause in the hardware itself. We ignore here permanent errors due to design errors or production issues. While hardware is designed with significant robustness margins, the processor will at runtime experience a constantly varying load. This has become even more important as modern processors and micro-controllers will have several blocks that can operate in parallel and access I/O and memory in parallel. If such a heavy load condition is encountered, even if only for a few clock pulses, short term peak current can be very high and the external as well as the internal circuits must be able to supply the current. It is very hard, if not impossible to generate test signals that can stimulate a processor in such a way that the worst case condition is created. Therefore, the only solution is to create a stress test application that is designed so that it exercises all blocks in parallel but with small time shifts so that when the test is run long enough, the condition is bound to happen. In our experience, using a multi-tasking RTOS is therefore also a

E. Verhulst et al., *Formal Development of a Network-Centric RTOS: Software Engineering for Reliable Embedded Systems*, DOI 10.1007/978-1-4419-9736-4_8, © Springer Science+Business Media, LLC 2011

good means of detecting lurking silicon 'bugs'. While such stress test are sometimes run for days, one should keep in mind that even one hour of testing for a 100 MHz processor produces a multiple of several trillion state transitions. If there is an error, it is bound to happen.

Assuming all such compiler and hardware issues have been tackled, the major metrics for measuring the quality of the final implementation of an RTOS are often expressed in terms of memory requirements, speed performance, real-time predictability as well as various other metrics like the quality of the code itself as this reflects how difficult it could be to port and maintain the RTOS. We discuss the main metrics and how they are related followed by measured data.

8.1.1 Code Size

Every processing system requires memory. Part of the memory will be used for program code (e.g. the RTOS kernel and the application tasks) whereas another part will be used to hold data. The latter can be local variables and datastructures but also data that is related to the application. Most processors today have a much higher clock speed than the bulk memory that is available; hence, processor manufacturers implement a memory hierarchy. The fastest memory are the internal registers. They can be read and written to at the speed of the processor (assuming each instruction takes one clock cycle). These registers might be complemented with some on-chip SRAM or a first level cache memory. Bigger processors will have second and third level caches and often external memory. Faster means more expensive and less density; hence, the external memory can be GBytes but operating with 100s of waitstates. All other types of memory are somewhere in between. For real-time systems this is not very good as the memory access times become statistical. If the code and the data is cached in level one cache, performance will be best. But as form time to time there will be cache misses, Worst Case Execution Times can be a few 100 times worse. Therefore, code size still matters as it will allow to use slower processors with less external memory and as a side-effect, also energy consumption will be reduced (Table 8.1).

In the project, we have in detail analysed the contribution of the various functional parts to the memory requirements. The processor was a small Melexis 16bit micro-controller, using a back-end port of the GNU compiler. This version of OpenComRTOS was still limited in functionality to prioritised task scheduling and port hubs (hence the prefic L0_). Using compile time switches variants were generated of the kernel to analyse the influence of leaving out more or less functionality. This resulted in code sizes ranging from 3,510 bytes to 904 bytes when using the -Os (compiler switch for obtaining smallest code size, often resulting in best performance as well) compiler option. The results are listed below, also showing the important influence of the compiler options used.

Table 8.1 Code size for OpenComRTOS kernel on MLX16

Target	GCC compiler	SP								MP					
MLX81001-27MHz	0.20 beta	Tiny		SendPacket–ReceicePacket+Start+small		All service, no WT		All services, NW, WT, Async		SendPacket–ReceicePacket+Start+small		All services no WT		All services NW, WT, Async	
Options		−O3	−Os	−O3	−Os	−O3	−Os	−O3	−Os	−O3	−Os	−O3	−Os	−O3	−Os
Kernal	code size (Bytes)	1,038	904	1,112	1,012	1,974	1,820	3,024	2,706	1,960	1,756	2,918	2,624	4,004	3,510
	readonly date (Bytes)	0	0	0	0	20	20	24	24	12	12	20	20	24	24
	Work space data size	20	20	26	26	40	40	52	52	34	34	46	46	58	58
Application with 2 Tasks and 2 semaphores	code size (Bytes)	258	144	354	202	354	202	354	202	354	202	354	202	354	202
	readonly date (Bytes)	0	0	0	0	0	0	0	0	0	0	0	0	0	0
	Work space date size	0	0	4	4	4	4	4	4	4	4	4	4	4	4
Total code size (Bytes)		1,425	1,120	1,643	1,315	2,505	2,123	3,555	3,009	2,491	2,059	3,449	2,927	4,535	3,813
Total static appl date size (bytes)		98	98	178	178	582	582	624	624	390	390	842	842	928	928
Data size (Bytes) per item	Port	2	2	6	6	6	6	6	6	6	6	6	6	6	6
	Packet	16	16	22	22	26	26	28	28	24	24	26	26	28	28
	Task	6	6	16	16	22	22	22	22	18	18	22	22	22	22
	Task input Ports	2	2	6	6	6	6	6	6	6	6	6	6	6	6
	Task context	0	0	0	0	6	6	6	6	0	0	6	6	6	6
	Timer					6	6	6	6			6	6	6	6
	Routing table									2	2	2	2	2	2

Notes:

1. MP = distributed, i.e. multi-processor (multi-node), version of OpenComR-TOS. All services are transparent of the topology and processor independent.
2. SP = single-processor (single-node) version of OpenComRTOS.
3. SPTiny = special derived variant of SP Small for MelexCM (MLX16x8) limited to 16 tasks (each with different priority) and 16 local ports.
4. Full = variant of OpenComRTOS supporting all services, i.e. currently:

   - L0_StartTask_W
   - L0_StopTask_W
   - L0_SuspendTask_W
   - L0_ResumeTask_W
   - L0_SendPacket_W, L0_SendPacket_NW, L0_SendPacket_WT, L0_Send-Packet_A
   - L0_ReceivePacket_W,   L0_ReceivePacket_NW,   L0_ReceivePacket_WT, L0_ReceivePacket_A
   - L0_AllocatePacket_W, L0_AllocatePacket_NW, L0_AllocatePacket_WT
   - L0_DeallocatePacket_W
   - L0_WaitForPacket_W, L0_WaitForPacket_NW, L0_WaitForPacket_WT
   - Note: the two phase services (L0_SID_A, and L0_WaitForPacket) were only implement in the SP variant.

5. Full, No WT = variant of OpenComRTOS supporting all services (see Full), except:

   - L0_SendPacket_WT
   - L0_ReceivePacket_W
   - L0_AllocatePacket_WT
   - L0_WaitForPacket_WT

6. Small = variant of OpenComRTOS, supporting:

   - L0_StartTask_W
   - L0_SendPacket_W
   - L0_ReceivePacket_W
   - Note: no support for Task arguments, using a restricted Task context or workspace, which may limit the debugging of tasks. (only StackPointer). Typically used with specific node configuration files eliminating non-essential static memory usage.

7. SPTiny = special derived variant of SP Small for MelexCM (MLX16x8) limited to 16 tasks (each with different priority) and 16 local ports.
8. Global ID size for TaskIDs and PortIDs:

   - MP Full, ML Small, SP Full: 16 bits (Local ID + Node ID, no SiteID nore ClusterID)
   - SP Small, SP Tiny: 8 bits (Local ID only)

9. Static allocated stack frames not included in Task context above (configurable per task).
10. Packet data size not included in above (compile-time configurable).
11. MP L0_StopTask_W does not include clean-up of state on remote nodes, only on local node.
12. Implementation of L0_EnterCriticalSection and L0_LeaveCriticalSection included as inline macros (mixed ASM and C).
13. Interrupt Service Routines (ISR) use a dedicated stack.

8.1.2 Total Memory Use

As memory requirements are determined not only by the code size, but also by the data requirements (static variables, stack space for each task, and global system datastructures), a further analysis was made to analyse the total memory requirements. Again the influence of the compiler options is important (Table 8.2).
Notes:

1. Measurements on MLX81001A TBC evaluation board configured without LIN driver.
2. MP Small, SP Small, and SP Tiny used with node configuration files eliminating non-essential static memory usage.
3. Packet data size of 8 bytes.
4. MP Full and SP Full include static kernel packet pool of 10 packets (280 bytes for above packet data size) (configurable).
5. MP Full and MP Small include static packet pool for Rx driver tasks of 5 packets. (140 bytes resp. 120 bytes for above packet data size) (configurable).

8.1.3 Influence of Processor Architecture

While above analysis gave indications on the relative contributions of selected functionalities and compiler options, by porting the RTOS to different processor architectures, one can also see the influence of processor instruction sets and in-struction length. The measurements were done using GNU-C as front-end compiler with target specific back-ends. The measurements were also done using a complete kernel with all hub functions to evaluate how each specific hub type adds additional memory requirements on top of the generic hub. The figures demonstrate that the architecture is indeed much more code size efficient. One can also see the influence of the instruction length (Table 8.3).

Table 8.2 Total memory used for OpenComRTOS on MLX16

/Bytes	Options	Kernel			Application			Total code size	Total static appl data size
		Code size	Read Only data	Data size	Code size	Read Only data	Data size		
MP Full	−Os	3,510	24	58	202	0	2	3,712	926
	0	4,004	24	58	354	0	2	4,358	926
Full, No WT	−Os	2,624	20	46	202	0	2	2,826	840
	0	2,918	20	46	354	0	2	3,272	840
Small	−Os	1,756	12	34	202	0	2	1,958	388
	0	1,960	12	34	354	0	2	2,314	388
SP Full	−Os	2,706	24	52	202	0	2	2,908	652
	0	3,024	24	52	354	0	2	3,378	652
Full, No WT	−Os	1,820	20	40	202	0	2	2,022	580
	0	1,974	20	40	354	0	2	2,328	580
Small	−Os	1,012	0	26	202	0	2	1,214	176
	0	1,112	0	26	354	0	2	1,466	176
Tiny	−Os	904	0	20	144	0	0	1,048	98
	0	1,038	0	20	258	0	0	1,296	98

Table 8.3 Code size figures (in 8 bit Bytes)

Service	MLX16	MB Xilinx	Leon3	ARM (M3)	XMOS
Hub shared	400	4,756	4,904	2,192	4,854
Port	4	8	8	4	4
Event	70	88	72	36	54
Semaphore	54	92	96	40	64
Resource	104	96	76	40	50
FIFO	232	356	332	140	222
Total	1,048	5,692	5,756	2,572	5,414

Table 8.4 Semaphore Loop times ($= 2$ signals, 2 tests, 4 context switches) in microseconds

	MLX16	MB Xilinx	Leon3[a]	ARM (M3)	XMOS
Clock (MHz)	6	100	40	50	100
	100.8	33.3	136.1	52.7	26.8

[a] Using external memory

8.1.4 Semaphore Loop

Another relevant test is called the semaphore loop test (explained in Appendix A). While simple, it is an adequate test as it measures more or less the minimum time needed to switch from one task to another. When done in a loop, it is also an adequate stress test for e.g. measuring interrupt latency (explained in Appendix A) because every context switch has to be atomically protected, hence interrupts are disabled. In the table below, we show the time taken for a full loop measures in microseconds but the reader should keep in mind the relative merit of the measured figures as they are dependent on the processor clock frequency and the access speed to program and data memory (Table 8.4).

8.1.5 Interrupt Latency

Another relevant test is called the interrupt latency test (explained in Appendix A). This test must be done with great care as it measures the reaction delay between a hardware event occuring (typically the raising of an interrupt by an external device, called IRQ below) and the moment the processor can start acting on it at the application level. Therefore, the measurement is processor specific but also application specific. Each application can have its interrupt system being set up differently, but the latency also depends on how the rest of the application is written (typically disabling interrupts because of context switches or the need for atomic access to certain hardware functions). Hence, the result of the test cannot be a single figure as some RTOS vendors proclaim.

What is important for a hard real-time system is that the interrupt latency remains below a reasonable low but strictly maximum value. In other words, when interrupt latency is measured, its histogram should a nice block diagram, but with a very high peak value at the beginning. For the test set-up, interrupt latency is the defined at two levels. The first level measures the time it takes from the hardware interrupt to the first "useful" instruction in the Interrupt Service Routine and the second level measures the time to the first "useful" statement in a waiting Task. As first useful statement we define the point where that data can be read. This measurement set-up is meaningful as it is realistic. Often hardware will trigger an interrupt related to data becoming available in a register. In the test set-up, a period hardware timer is used to create these conditions. The data will not remain forever in this register as it will be overwritten at the next interrupt (or sometimes even sooner if the hardware was not designed for it). Hence it is important to know how much time an application running an RTOS on a specific processor needs to safely read the data. This also determines the maximum "sampling" rate for a particular I/O channel. In order to make the test even more realistic, a worst case small application is running in the background. We used here the semaphore look test described in the previous paragraph as it results in continuous context switches.

The measured IRQ to ISR latency on ARM Cortex M3 50 MHz are as follows:

- Minimal: 300 ns
- Maximal: 2,140 ns
- 50% of samples less or equal: 400 ns

The measured IRQ to Task latency on ARM Cortex M3 50 MHz are as follows:

- Minimal: 12 μs
- Maximal: 25 μs
- 50% of samples less or equal: 17 μs

The reader can verify these figures and the resulting histogram in Appendix A.

Part IV
Appendix

Appendix A
OpenComRTOS-Suite 1.3 Usage Tutorial

The previous chapters of this book concentrated the theoretical foundations of OpenComRTOS and the formal modelling effort that was done in order to develop a trustworthy Real Time Operating System (RTOS). This chapter concentrates on the development of applications which utilise OpenComRTOS, as provided by OpenComRTOS-Suite 1.3. It is designed as a tutorial the reader can follow step by step. Section A.1 gives a detailed introduction to OpenVE (the main development tool of the OpenComRTOS-Suite) to develop a Semaphore-loop executing on a single win32-node. The resulting project is then extended in Sect. A.2, to utilise multiple nodes (MP), by introducing a second win32-node and connecting the two win32-nodes using TCP/IP. This illustrates how easy it is to construct distributed systems using the OpenComRTOS-Suite. One important aspect of the system development process is validating that the system operates as expected, especially with respect to timing constraints. For this purpose, OpenComRTOS provides a tracing mode which gathers the execution trace of a node, and the OpenTracer tool is used to display the gathered execution traces. Section A.3 illustrates how to gather an execution trace from an embedded ARM-Cortec-M3 node, save it on a win32-node for examination, and then display it using OpenTracer. The last tutorial in this chapter in Sect. A.4, illustrates how to measure the Interrupt Request (IRQ) to Interrupt Service Routine (ISR) and IRQ to task latencies of an ARM-Cortex-M3 node and display them using a native win32 application.

The interested reader can freely download a win32 version of OpenComRTOS from the altreonic website.

A.1 Developing a Single Node Semaphore-Loop Project

This section guides the reader step by step on how to develop a Semaphore-Loop executing on a single win32-node, using OpenComRTOS-Suite 1.3.

E. Verhulst et al., *Formal Development of a Network-Centric RTOS: Software Engineering for Reliable Embedded Systems*, DOI 10.1007/978-1-4419-9736-4_9, © Springer Science+Business Media, LLC 2011

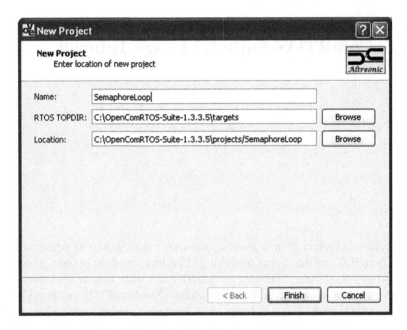

Fig. A.1 Screenshot of OpenVE's the 'New Project' dialogue

1. Creating a new Project:

   a. Click on the entry 'New Project' from the 'File' Menu.
   b. The 'New Project' dialogue will open, see Fig. A.1. This dialogue defines the following project wide settings:

      • Name: The name of the project, in this example 'SemaphoreLoop' was chosen as project name.
      • RTOS TOPDIR: This option defines which OpenComRTOS Kernel-Image should be used for this project.
      • Location: The directory in which the project should be created in.

      For 'RTOS TOPDIR' and 'Location' OpenVE selects useful default values which should not be changed at this point in time.
   c. Once all settings are as desired click on the button labelled 'Finish', to create the new project.

2. Create a new Win32 Node: After creating the new project, OpenVE automatically opens the Topology view, where the user can create new nodes to represent the topology. To create a new Win32-Node perform the following operations:

   a. Click on the icon labelled 'win32', see Fig. A.2, and then onto the canvas.
   b. This opens up the dialogue shown in Fig. A.3, where we will give the node the name 'Win32_Node'.

   Figure A.4 shows the resulting topology with one a node called 'Win32_Node'.

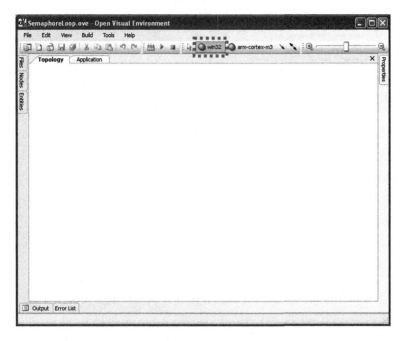

Fig. A.2 OpenVE with opened Topology View (no nodes defined yet)

Fig. A.3 The dialogue to specify the properties of the new win32-node

3. Create two Tasks: Task1 and Task2: Tasks are elements of the 'Application Diagram' thus we have to switch to the Application view of the project, using the tabs on top of the canvas. To create a new task click on the icon symbolising a Task, see Fig. A.5, and then click on the canvas. The 'New Task' dialogue shown in Fig. A.6 will appear and ask you to specify the following things:

- Name: The name of the new Task, fill in 'Task1'.
- Node: On which Node should this Task run, in our case there is only one node, thus select 'Win32_Node'.

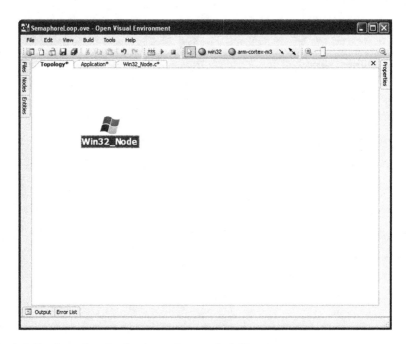

Fig. A.4 Topology view showing the newly created win32-node

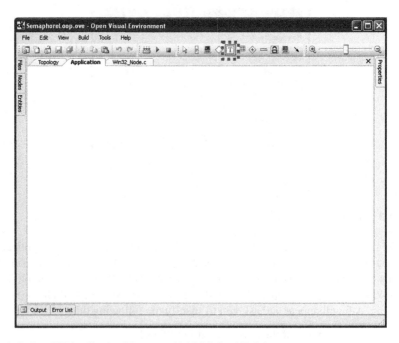

Fig. A.5 OpenVE Application Diagram with highlighted Task button

Fig. A.6 'New Task' dialogue, with highlighted 'Task Entrypoint' creation button

Fig. A.7 The Task Entrypoint creation dialogue, showing the source code that will be generated

- Stack Size: How many bytes of stack does the new Task have. On win32 technically not necessary to fill in, but it is custom to give a task 1,024 bytes.
- Task Entry Point: The function that represents the new Task. Add a new one here using the button with the plus sign. This will open the 'New Entry Point' dialogue, shown by Fig. A.7. Call the new entry point 'Task1_EP'.

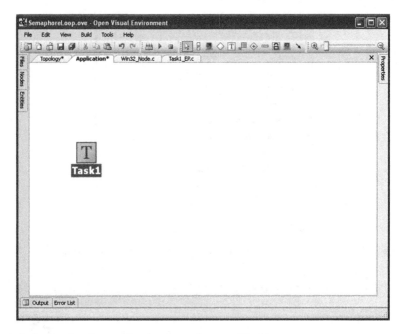

Fig. A.8 Application diagram showing the newly created Task1

After acknowledging everything the Application Diagram of your project should look similar to the one shown in Fig. A.8. Now repeat these steps to create Task2. Figure A.9 shows the resulting Application diagram.

4. Create two Semaphore Hubs: Sema1 and Sema2 To create a new Semaphore Hub click on the Semaphore icon, marked in Fig. A.9, and then click on the canvas. OpenVE then presents the 'New Semaphore' dialogue, shown in Fig. A.10. Similar to a Task a Hub must be mapped onto a node, also it must be given a name.

5. Establish the following interactions:

   a. Task1 signals Sema1, using '_W' semantics (`L1_SignalSemaphore_W(Sema1)`);
   b. Task2 tests Sema1, using '_W' semantics (`L1_TestSemaphore_W(Sema1)`);
   c. Task2 signals Sema2, using '_W' semantics;
   d. Task1 tests Sema2, using '_W' semantics;

To create an interaction select the interaction symbol and then draw a line either from a Task to a Hub (put interactions, in this example `L1_SignalSemaphore_W(Sema1)`) or from a Hub to a Task (get interactions, `L1_TestSemaphore_W(Sema1)`). Upon releasing the left mouse button OpenVE presents a properties menu from which the valid interactions can be chosen (interaction selection menu). An example of this menu is shown in Fig. A.11. The resulting line of source code gets immediately inserted in the corresponding Task Entry Point. In the end the Application Diagram should look similar to the one shown in Fig. A.12.

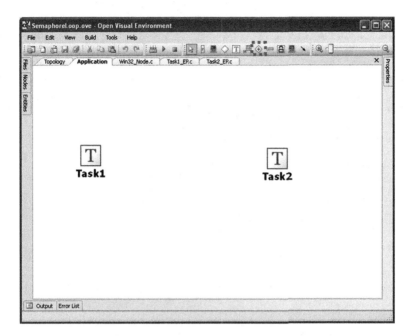

Fig. A.9 Application diagram showing both Task1 and Task2

Fig. A.10 The 'New Semaphore' dialogue of OpenVE

6. Move the interactions into the while-loop in the source code of Task1_EP and Task2_EP. Figure A.13 marks the two interactions to be moved into the while loop.
7. Add an Stdio Host Server to the System, and adjust the Task Entrypoints to print messages after every iteration of their loop. Adding StdioHostService is similar to adding a Semaphore Hub, Fig. A.14, shows the corresponding dialogue.

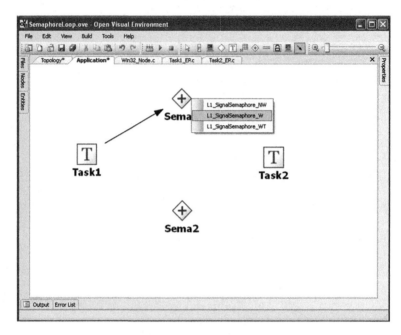

Fig. A.11 Application diagram with all entities, showing the interaction selection menu

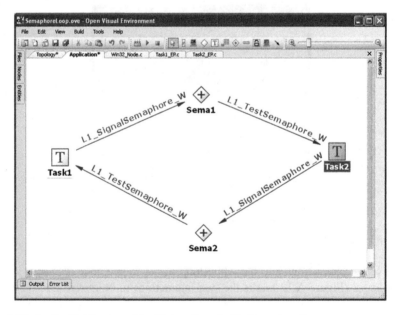

Fig. A.12 Application diagram with all Interactions for the Semaphore Loop

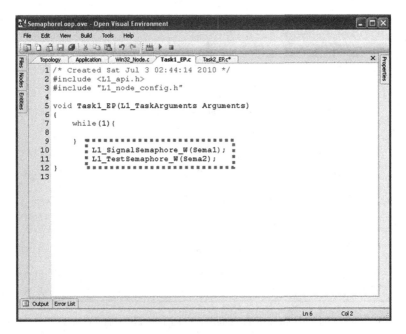

Fig. A.13 Source code for Task1, the incorrectly placed interactions highlighted

Fig. A.14 The 'New stdioHostServer' dialogue of OpenVE

Map the Stdio Host Server onto the Win32_Node and give it the name 'Shs'. Now add the interaction 'Shs_putString()' to both Task Entrypoints, see the resulting Application diagram in Fig. A.15. Let each of the task print a newline terminated string onto the console, using the Stdio Host Server, Fig. A.16 shows the resulting code for Task1_EP.

8. Build and Run the project as you did in the previous section. If everything is OK, you should see a console output similar to the one shown in Fig. A.17.

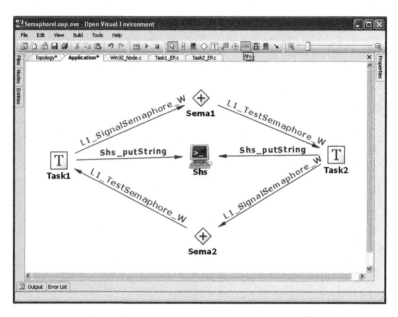

Fig. A.15 Application diagram with the complete Semaphore-Loop and the Stdio Host Server

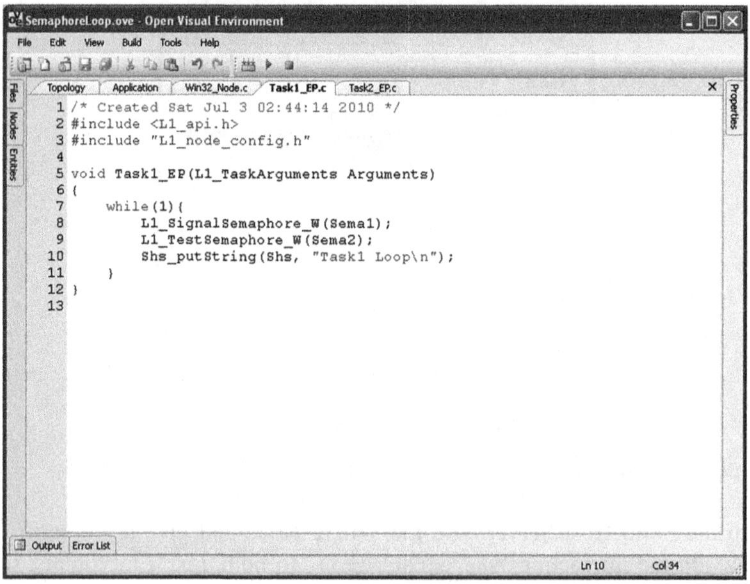

```
 1 /* Created Sat Jul 3 02:44:14 2010 */
 2 #include <L1_api.h>
 3 #include "L1_node_config.h"
 4
 5 void Task1_EP(L1_TaskArguments Arguments)
 6 {
 7     while(1){
 8         L1_SignalSemaphore_W(Sema1);
 9         L1_TestSemaphore_W(Sema2);
10         Shs_putString(Shs, "Task1 Loop\n");
11     }
12 }
13
```

Fig. A.16 Source code of Task1 with Semaphore and Stdio Host Server Interactions

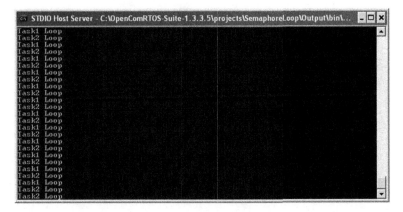

Fig. A.17 Console output upon running the 'SemaphoreLoop' project

A.2 Going Distributed with OpenComRTOS

This section turns the system developed in the previous section into a distributed system. To achieve this, follow these steps:

1. Add another Win32 Node to the topology of the project and call it 'Win32_Node2'.
2. To establish a communication link between the two nodes, they must be configured to offer a Link-Port. Add to each of the nodes an TCP-IP Link-Port following these steps:

   a. Right click on the node to open its properties menu.
   b. In the properties menu select the entry called 'Edit Link Ports'. This will open the Link-Port editing dialogue, shown in Fig. A.18.
   c. In this dialogue select the Link-Port type 'tcp' and press the 'Add Link Port' button (the button labelled '+', highlighted in Fig. A.18) to add the Link-Port. Give the Link-Port a name, the other settings chosen for this Link-Port should not be changed.
   d. Press OK to close the dialogue.

3. Now the Link between the two Nodes must be established, this is done by selecting the 'bidirectional Link icon' (the arrow with arrow heads at both ends), and drawing a line between the two nodes. OpenVE will now present the link configuration dialogue, pictured in Fig. A.19. There you select the two Link-Ports and press OK to create the Link. Figure A.20 shows the resulting Topology Diagram, with both Nodes linked.
4. To make this a real distributed system it is necessary to map at least one of the Application Diagram entities onto the newly created node, in this example

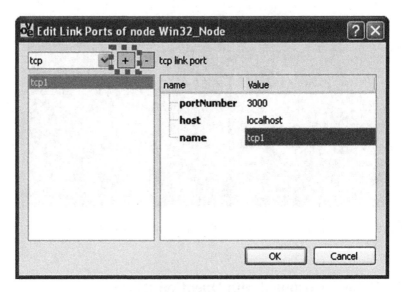

Fig. A.18   Edit Link Ports Dialogue, with highlighted 'Add Link Port' button

Fig. A.19   OpenVE link configuration dialogue

I chose to map Task2 and Sema2 onto the newly created Node 'Win32_Node2'. To do this follow these steps:

a. Change to the Application Diagram.
b. Open the 'Properties' side-pane on the right hand side and pin it, so it stays open permanently.

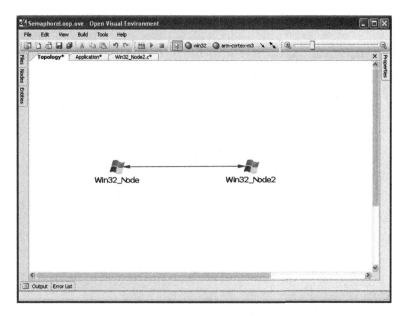

Fig. A.20 Topology of the two Win32 Nodes connected with a bidirectional Link

   c. Select the entity to remap. The 'Properties' side-pane will now show all the properties of the entity, including the property 'node'. See Fig. A.21 for a screen-shot of OpenVE with opened 'Properties' side-pane.

   d. In the combo-box for 'node' select the Node 'Win32_Node2'.

Repeat the last two steps for each entity that should be remapped.

5. The system is now ready to be rebuilt and run, and 'Win32_Node' will provide the same console output as previously, shown in Fig. A.17. This is caused by the fact that OpenComRTOS has been designed for exactly these systems and great care has been taken to make it scalable. This means that the logical behaviour does not change independently of whether the system consists of only one node or 1,000 nodes. The only thing that will change is the execution speed, caused by the fact that communication links always introduce a certain latency, and there is nothing we can do about this, except optimising the hardware and the drivers.

In contrast 'Win32_Node2' does only show some Link debugging messages, see Fig. A.22 for a screen shot of the console outputs of both Win32_Node and Win32_Node2.

A.3   Tracing in OpenComRTOS

OpenComRTOS allows the user to examine which operations it performed at what time. This is called tracing. Naturally, not every instruction can be traced, and this is also not necessary to understand of what has happened in the system.

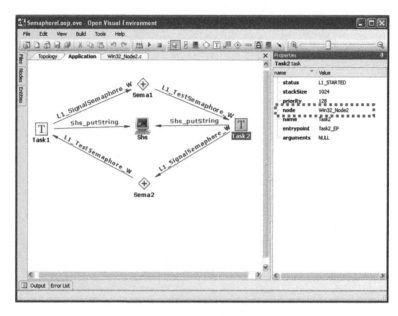

Fig. A.21 OpenVE with open 'Properties' side-pane and highlighted 'node' property

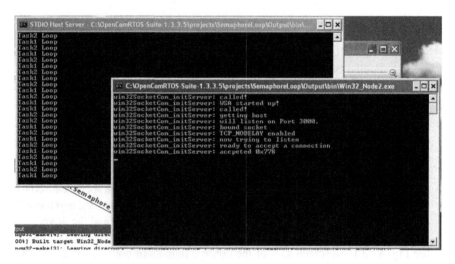

Fig. A.22 Console output of both win32-nodes

The OpenComRTOS kernel has a tracing mode, in which it collects the following events:

- Scheduling Events – which task ran at what time.
- Service Requests by Tasks – at what time did a task issue a specific service request. Naturally, the nature of the service request is captured as well.

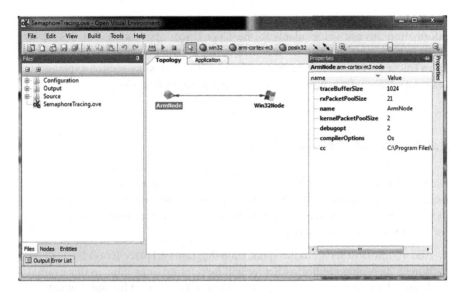

Fig. A.23 OpenVE with open Property Pane

- Hub interactions – when did the kernel perform an interaction with a hub.
- Node interactions – these information are used by the tracer to combine multiple traces that represent (parts) of a multi node system.

A.3.1   How to Enable Tracing

To enable the tracing mode you have to set node specific properties (node-properties). To set a node-property, open the node-property pane by double clicking on a node in the topology-diagram, your OpenVE window should now look similar to Fig. A.23.

There are two node-properties relevant to tracing:

- `debugopt` must be set to 1 or 2 – `debugopt` defines the debug-mode a node runs in. The property `debugopt` may have the following values:
    - 0: tracing disabled – no trace information gets recorded at all.
    - 1: limited tracing mode – all trace information except 'Service Request by Tasks' gets recorded. This is used to reduce the overhead caused by the tracing operations.
    - 2: full tracing mode – all trace information gets generated.
- `traceBufferSize` – `traceBufferSize` defines how many past events get recorded on a particular node. It defaults to '1024', its upper limit is defined by the amount of memory available on the Node.

The Node now collects trace-information in its trace-buffer, but these trace information are not yet available to the OpenTracer application. For this, it first needs to be written to a file, i.e. the trace-buffer needs to be dumped, only then, the OpenTracer application can interpret the trace. The following section explains how to retrieve trace information from a Node to generate a trace-file.

A.3.2 How to Retrieve a Trace

An embedded Node has usually no file system available which could be used to store a trace. Instead, OpenComRTOS Nodes can transfer the contents of the trace-buffer to a StdioHostServer which will then write the retrieved trace information into a file for the OpenTracer application.

1. Add a StdioHostServer to the application diagram and place it on a Node of type Win32. The added StdioHostServer will be referred to as 'Shs' in this example. A StdioHostServer is a task which offers a range of stdio functionalities to embedded Nodes, such as the ARM Node. One of which is to receive the contents of a trace-buffer and write it onto a disk.
2. Add the instruction DumpTraceBuffer(Shs) to one of the tasks. This is the actual instruction which will transfer the contents of the trace-buffer to the StdioHostServer with the name 'Shs'. The retrieved trace information are then written to a file with the extension 'trace1' (in the following this file will be referred to as trace1-file).

A.3.2.1 Extending the Semaphore Example with Tracing

The project 'Semaphore_Tracing' is a tracing enabled version of the previously shown Semaphore-Example. In addition to the changes explained in Sects. A.3.1 and A.3.2, the following has been changed:

1. All Tasks and Hubs of the Semaphore Loop have been mapped onto the ARM node.
2. In one of the two tasks of the Semaphore-Loop a for-loop which lets the loop execute for 500 times, has been added
3. Before writing out the trace-buffer a message is displayed on the StdioHostServer announcing that the dumping of the trace takes place.
4. The task calls the function DumpTraceBuffer() to write the contents of the trace buffer into a file;
5. After the dumping of the trace is completed another message is displayed which asks the user to press enter to continue the execution. This is necessary because otherwise the system would continuously overwrite the trace file it generated.

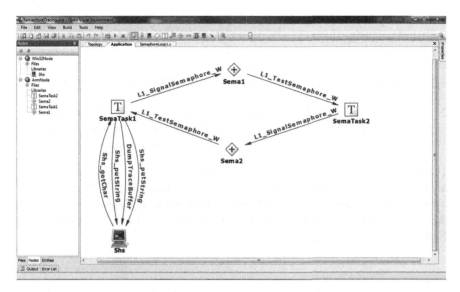

Fig. A.24 Tracing enabled Application Diagram

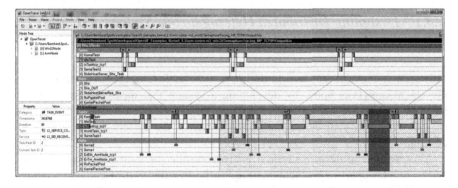

Fig. A.25 OpenTracer displaying parts of the `SemaphoreTracing_MP_TCPIP` example trace

Depending on your application you may need to perform similar changes. Figure A.24 shows the resulting Application Diagram for the Semaphore Example.

Execute the generated binaries. To display the retrieved trace open the trace-file (extension 'trace') with OpenTracer, this should give you an output similar to the one shown in Fig. A.25.

A.3.3 Retrieving and Displaying Traces from Distributed Systems

The natural habitat of OpenComRTOS are distributed heterogeneous systems. This section will explain the steps to display traces from such systems. The procedure for acquiring the trace buffer contents from multiple Nodes simultaneously is similar to acquiring the trace information from a single Node. Thus the steps are similar:

1. Add a StdioHostServer `shs` to the application diagram and place it on a Node of type Win32.
2. Add the instruction `DumpTraceBuffer(Shs)` in at least one Task per Node. The tricky bit is that one must ensure that all nodes dump almost simultaneously, otherwise the tracer cannot interconnect the collect trace information.

The Example: `SemaphoreTracing_MP_TCPIP` demonstrate the collection and dumping of trace information for a distributed heterogeneous system. Figure A.25, shows a part of the expected display from OpenTracer when opening both Trace files. Note the red lines connecting both traces, these mark the way a Packet travels from one Node to another.

A.4   Measuring the Interrupt Latency of OpenComRTOS

A.4.1   Designing Distributed Heterogeneous Systems Using the OpenComRTOS Suite

This section demonstrates how to design distributed heterogeneous systems using the OpenComRTOS Suite, by developing a system that measures the interrupt latencies of an ARM-Cortex-M3 running OpenComRTOS. Before starting this it is first necessary to define the requirements of the system we are trying to develop.

Most peripheral devices inside a micro controller or even a standard PC use hardware interrupts to signal that they require the attention of the Central Processing Unit (CPU), this is called the devices send an IRQ. Examples of IRQs are: the user pressed a button, the Universal Asynchronous Receiver/Transmitter (UART) RX-buffer is full, the UART TX-buffer is empty, a timer has expired, and so on. Once the CPU has detected the occurrence of an interrupt it stops what is currently doing and instead executes the code that handles the interrupt, afterwards execution resumes where the CPU was interrupted. Due to this it is the custom for most Operating Systems to disable the handling of interrupts when modifying global data structures, the CPU is then said to be in a critical section. This means that if an IRQ occurs while the CPU is in a critical section the interrupt will only be handled after the critical section has been completed. Depending on the length of the critical section quite some time might expire between the occurrence of an IRQ and its handling by the CPU.

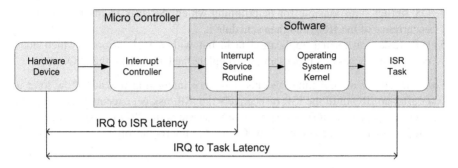

Fig. A.26   Stages of IRQ handling in a typical Microcontroller System

A.4.1.1   How Micro Controllers Handle Interrupts

This section gives background information on how IRQ handling takes place on standard micro-controllers that runs an RTOS. Figure A.26, shows the hardware and software entities involved from generating an IRQ in a hardware device, to signal the occurrence of a hardware event, to the handling of the IRQ in an application task, i.e. a task provided by the developer of the application that runs on top of the RTOS. There are a number of individual stages involved:

1. Hardware Device – which has previously been programmed to generate an IRQ. This usually happens by asserting a pin on the CPU.
2. Interrupt Controller – keeps track of pending IRQ and whether or not interrupts are permitted at the moment. Many interrupt controllers allow the CPU fine grain control which IRQs it permits and which not, this is commonly referred to as interrupt masking. Interrupts that cannot be masked, typically the timer interrupt, are referred to as Non Maskable Interrupts (NMIs). In case of multiple pending IRQs the interrupt controller determines which to signal first to the CPU. Interrupt controllers can be inside the CPU or outside the CPU. Most micro controllers have the interrupt controller built in, because it reduces the component count and makes developing software easier. One exception of this is the Xilinx Microblaze Softcore, which only offers one interrupt pin and requires for anything more an external interrupt controller. This setup requires interrupt dispatching (which IRQ occurred, and which ISR to call) which increases the interrupt latency drastically.
3. Interrupt Service Routine – the piece of code that the CPU executes once it has been interrupted. Usually, the developer registers a separate ISR for each IRQ. ISRs are meant to be as short as possible, this means that any complex processing is deferred to a later point in time, outside an ISR. The ISR itself only does the bare minimum to ensure that the hardware device can continue its operation. This means that the ISR signals the Operating System (OS) kernel about the occurrence of the IRQ. For this purpose it is custom to immediately schedule the OS kernel after an ISR occurred.

4. OS Kernel – responsible for determining which ISR task to inform about the occurrence of the IRQ, and then schedule it.
5. ISR Task – a normal application task, but with a high priority, which does perform the necessary processing to complete the handling of the IRQ.

For a developer of real time systems it is very interesting to know how long it takes after an IRQ until the ISR respectively the ISR task get executed. This section develops an OpenComRTOS system which measures the IRQ to ISR respectively the IRQ to Task latencies on an ARM Cortex M3 microcontroller.

A.4.2   Presenting the Measurement Results

The previous section defined what we want to measure. The interrupt latencies depend on whether or not the CPU was inside a critical section at the time the hardware device issued the IRQ. Thus, there is not a single figure, but a range of latencies and how often they occurred. Therefore, we will have to perform a statistical analysis of the measured data, and present it in from of a histogram to the user. This kind of display operation requires a decent display, something an embedded micro-controllers and their evaluation/development kits usually does not provide.

A.4.2.1   Requirements

Our analysis has determined the following requirements:

• R1: Measurement application running on an ARM-Cortex-M3 micro controller, which is able to measure the following:

  – IRQ to ISR Latency.
  – IRQ to Task Latency.

• R2: GUI Application for statistical analysis and histogram display.
• R3: Communication between the micro controller and the GUI Application.

A.4.3   Specifying the System

The following give the specification of the system, for each requirement there will be a sub specification.

A.4.3.1   S1: The Specification for R1

The following list represents the specification S1, which represents requirement R1:

- An OpenComRTOS ARM-Cortex-M3 node: ArmNode.
- Automatic Reload Counter to be used as IRQ Source. An automatic reload counter is available in almost all micro controllers to be used as a periodic timer. The one on the chosen ARM-Cortex-M3, counts backwards from a predefined value towards zero. Once it reaches zero, it generates an IRQ to inform the CPU about this and then reloads a predefined value. In OpenComRTOS we use it to generate a 1 ms periodic timer tick.
- Automatic Reload Counter ISR to be modified to do the following:

  - Store the value of the Automatic Reload Counter in a global Variable once it has been started. This is the IRQ to ISR Latency.
  - Signal an OpenComRTOS Event using '_NW' semantics.

- A high priority task to represent the ISR task, which does the following:

  - Waits for an OpenComRTOS Event to become signalled and then acquires the current value of the Automatic Reload Counter in a local variable, this is then the IRQ to Task Latency.
  - Stores the previously measured IRQ to ISR Latency in a local variable.
  - Send the collected latencies to an OpenComRTOS Port using '_W' semantics (Communicating Sequential Processes (CSP)-Channel).

- The CPU executes a Semaphore Loop to ensure that it performs constant context switching. The semaphore loop consists of two tasks and two OpenComRTOS Semaphores, which are connected in such a way to achieve constant context switching between the two tasks.

A.4.3.2   S2: The Specification for R2

The following list represents the specification S2, which is a refinement of R2:

- An application that displays a histogram for each of the measured latencies, later on referred to as 'Latency-GUI'.
- The Latency-GUI performs the following statistical analysis, for each of the latencies:

  - Minimal Latency
  - Maximal Latency
  - 50% Latency – this is the latency below which 50% of all measured latencies are.

- The Latency-GUI application is a normal MS-Windows program written in C++ using QT. Figure A.27 shows the user interface of this application.

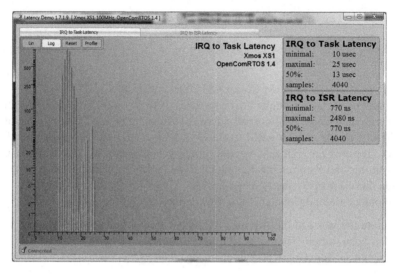

Fig. A.27   Screenshot of the Interrupt Latency GUI Application

- Measurements get send to the Application using a local TCP-IP connection. The Latency-GUI listens on port 4004 for connections from a local application.

A.4.3.3   S3: The Specification for R3

The following list represents the specification S3, which refines requirement R3:

- An OpenComRTOS Win32 Node: Win32Node.
- This node connects with the ArmNode using a TCP-IP link.
- The Win32Node contains a port which is used to exchange data between the Collector_Task and the Receiver_Task.

A.4.4   Implementation

A.4.4.1   Topology

Based on the specifications we can now develop the topology. There are three different entities in our Topology:

- ArmNode: This entity performs the measurements. It is connected over an OpenComRTOS TCP-IP link with the Win32Node.

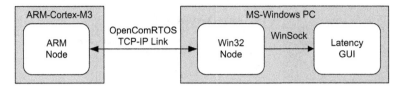

Fig. A.28 Interrupt latency measurement system topology

- Win32Node: This entity acts as interface between the ArmNode and the Latency GUI application.
- Latency GUI application: This acts as a data sink for the measurement data and displays them.

Figure A.28 shows the graphical representation of the topology.

A.4.5   Application

The application consists of the following entities:

- Entities related to taking and processing the measurement:

  - Automatic Reload Counter ISR – takes the IRQ to ISR latency measurement and signals the ISR_Event.
  - ISR_Event – synchronises the ISR and the Collector_Task.
  - Collector_Task – waits for the ISR_Event to become signalled and then takes the IRQ to Task latency measurement. Send the collected measurement data to the Data_Port.
  - Data_Port – acts as data exchange and synchronisation primitive between the Collector_Task and the Receiver_Task.
  - Receiver_Task – waits for a packet to become available on the Data_Port, and then sends the resulting data to the Latency GUI application.

    - SemaTask1 – executes the following in a loop: signal Sema1, then test Sema2
    - SemaTask2 – executes the following in a loop: test Sema1, then signal Sema2
    - Sema2 – synchronisation of SemaTask1 and SemaTask2
    - Sema1 – synchronisation of SemaTask1 and SemaTask2

Figure A.29 shows the resulting Application diagram and the mapping of the entities onto the different OpenComRTOS Nodes.

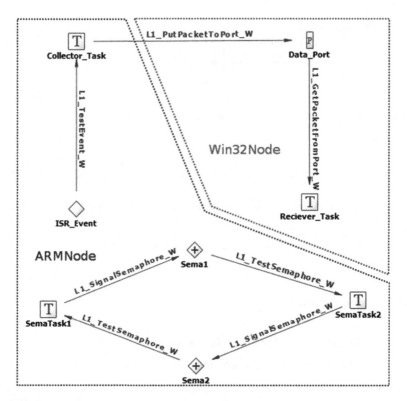

Fig. A.29 Interrupt latency measurement system application diagram

A.4.6   Collected Measurement Results

Figure A.30 shows the collected measurements for the IRQ to ISR latency:

- Minimal: 300 ns
- Maximal: 2,140 ns
- 50% of samples less or equal: 400 ns

Figure A.31 shows the collected measurements for the IRQ to task latency:

- Minimal: 12 μs
- Maximal: 25 μs
- 50% of samples less or equal: 17 μs

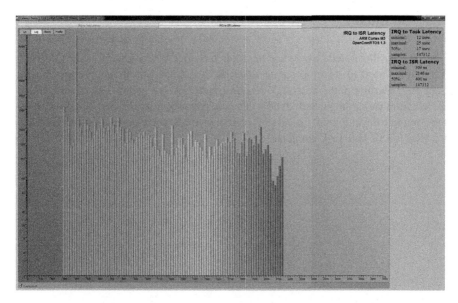

Fig. A.30 Measured IRQ to ISR Latency on ARM Cortex M3 50MHz (logarithmic scale)

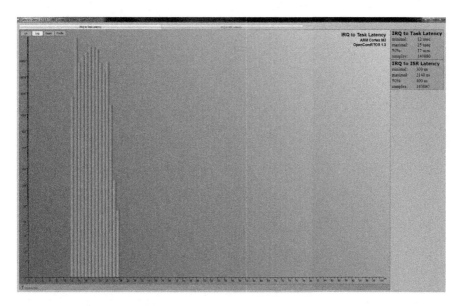

Fig. A.31 Measured IRQ to Task Latency on ARM Cortex M3 50MHz (logarithmic scale)

A.5   Summary

This chapter illustrated how to use the tools provided by the OpenComRTOS-Suite 1.3 can be used to develop single node, Sect. A.1, and multi node applications, Sect. A.2. Furthermore, Sect. A.3 demonstrated how to gather and display trace information using the provided tools. The chapter closed with Sect. A.4 illustrating the complete development process of a distributed heterogeneous system to measure the IRQ to ISR and Task latencies.

Appendix B
Foundations for TLA$^+$ and Temporal Logic

B.1 Introduction

B.1.1 Goal: Increased Awareness in Specifying Systems

As mentioned in Chap. 4, mathematics is the main intellectual tool for the genuine systems designer. In particular, the user of automated model or proof checkers can benefit from mathematical reasoning about both the system descriptions (e.g. programs) and the conditions (e.g. invariants, temporal formulas) in various respects. Here we mention only a few.

a. The link between informal specifications and formal descriptions, often considered the weak link, can be considerably strengthened by formalizing different informal views and then formally exploring their relationships (Boute 2006a,b). The formalizations can even be (advantageously) in different notations, captured by a unifying framework (Boute 2005) for the purpose of reasoning (e.g. about correspondences and disparities).

b. More specifically, in model checking the conditions to be met are usually expressed by temporal formulas, but these are often unintuitive and even confusing, as also observed by Lamport (2002a). This difficulty can be alleviated by *patterns* (Dwyer et al. 1998, 1999; Dwyer and Hatcliff 2002), i.e. fixed temporal formulas "known" to express some property of interest. Such an approach is used in the Bandera project (Ban 2003; Dwyer and Hatcliff 2002) for model checking concurrent Java software. However, there is no "complete" collection of patterns, and the specifier/designer must be able to understand existing patterns and design new ones. Clarifying the intuitive meaning of patters, exploring the relationship between them and designing new ones, is again most conveniently done via mathematics (Boute and Verlinde 2003).

E. Verhulst et al., *Formal Development of a Network-Centric RTOS: Software Engineering for Reliable Embedded Systems*, DOI 10.1007/978-1-4419-9736-4_10,
© Springer Science+Business Media, LLC 2011

So, rather than "hiding the math", we aim at making the math very accessible, bringing the opportunities just outlined within easier reach of the specifier/ designer.

B.1.2   Approach and Overview

Chapter 4 provided an introduction-by-example to TLA$^+$, illustrating how basic mathematics constitutes a flexible specification language. Examples were taken from the RTOS design project.

This appendix supplies a mathematical basis for readers interested in going beyond just *reading* specifications, but also want to *build* them. To achieve this, we follow a general and very efficient approach. An introduction to TLA$^+$ presenting its many constructs separately would require several dozens of pages, and can be found in Lamport's book (Lamport 2002a). Here, we use a generic formalism that is designed orthogonally and can be covered in a few pages. It can then be used for presenting other formalisms simply and succinctly, as will be done for TLA$^+$ in a few tables. It also supports *calculational reasoning* about formal specifications.

A related formalism that turned out to offer the required expressivity and reasoning power in general is a functional predicate calculus (Boute 2005). It allows engineers to calculate with predicates and quantifiers ($\forall$, $\exists$) as fluently as they were taught to do with derivatives and integrals. In other words, reasoning is calculational, as also advocated by Dijkstra (1990), Gries (1991), Gries and Schneider (1993) and others.

This appendix achieves similar benefits for specification/design by an approach that is formal, yet very accessible and suitable as a tool for discovery and developing intuition. It is organized as follows.

Section B.2 briefly presents a unifying formalism usable for reasoning about specifications, views and designs, and for capturing various tool-supported languages. After half a dozen pages of basic mathematics and some conventions for operator design clarified by examples, this formalism enables covering TLA$^+$ in two convenient tables.

In view of the aforesaid goal (a), Sect. B.3 discusses style issues in faithfully reflecting informal specifications and arguments in a formal way, together with illustrations.

In view of goal (b), Sect. B.4 is devoted to deriving a temporal calculus, establishing a convenient style of derivation, and showing applications in reasoning about temporal formulas expressing liveness issues in TLA$^+$. The power of the approach is illustrated by calculational derivations that are much simpler than classical proofs (Lamport 2002a), and often discover stronger results without knowing them in advance.

B.2   A Unifying Formalism

B.2.1   Rationale

A formal mathematical language is valuable insofar as it supports the design of precise calculation rules that are convenient in everyday practice.

In this sense, common mathematical conventions are strong in Algebra and Analysis (e.g. rules for $\int$ in every introductory Analysis text), weaker in Discrete Mathematics (e.g. rules for $\sum$ only in very few texts), and poor in Predicate Logic (e.g. disparate conventions for $\forall$ and $\exists$, rules in most logic texts not suited for practice). This is reflected in the degree to which everyday calculation in the respective areas can be called "formal", and inversely proportional to the needs in Computing Science.

Entirely deficient are the conventions for set comprehension. Common expressions such as

$$\{m \in \mathbb{N} \mid m < n\} \quad \text{and} \quad \{2 \cdot n \mid n \in \mathbb{Z}\}$$

may look innocuous, but exposing their structure as

$$\{v \in X \mid p\} \quad \text{and} \quad \{e \mid v \in X\}$$

(with the metavariables below) reveals the ambiguity: the example

$$\{n \in \mathbb{N} \mid n \in \mathbb{Z}\}$$

matches both. Calculation rules are nonexistent.

Funmath (Functional Mathematics) is not "yet another computer language" but an approach to structure formalisms by conceiving mathematical objects as functions whenever convenient – which is quite more often than common practice reflects. Four constructs suffice for synthesizing most (all?) common conventions without their ambiguities and inconsistencies, and also yield new yet useful new forms of expression, such as point-free expressions. This section summarizes only the syntax and main definitions; the calculation rules are treated extensively in (Boute 2005).

B.2.2   Syntax

To facilitate adoption of this design in other formalisms, we avoid a formal grammar. Instead, we use the following metavariables: i for a (tuple of) identifiers, and for expressions: v, w: (tuple of) variable(s); d, e: arbitrary; p, q, r: boolean; X, Y: set; f, g: function; P, Q: predicate; F, G: family of functions; S, T: family of sets. By "family of X" we mean "X-valued function". Here are the four constructs.

1. An *identifier* can be any (string of) symbol(s) except markers (binding colon and filter mark, abstraction dot), parentheses ( ), and a few keywords (**def, spec**). Identifiers are *introduced* by *bindings*

$$i : X \wedge p,$$

read "*i* in *X* satisfying *p*". The *filter* $\wedge$ *p* (or **with** *p*) is optional, e.g.

$$n : \mathbb{N} \quad \text{and} \quad n : \mathbb{Z} \wedge n \geq 0$$

are interchangeable.

Definitions, of the form

def *binding*,

introduce *constants*, with global scope. Existence and uniqueness are proof obligations. This is not the case for *specifications*, of the form

spec *binding*.

Example:

def *roto* $: \mathbb{R}_{\geq 0}$ **with** $roto^2 = 2$.

Well-established symbols (e.g. $\mathbb{B}, \Rightarrow, \mathbb{R}, +, \sqrt{\ }$) are seen as predefined constants.

2. A *function application* has the form $f\ e$ in the default *prefix* syntax. A function identifier is called an *operator* and clearly must not be confused with a TLA$^+$ operator. For an operator other affix conventions can be specified by dashes in its binding, e.g. $-\star- : x^2 \rightarrow X$ for infix. Prefix has precedence over infix. Parentheses are used for overriding precedence rules, *never* as an operator. Application may be partial: if $\star$ is an infix operator, then $(a\star)$ and $(\star b)$ satisfy

$$(a\star)\, b = a \star b = (\star b)\, a.$$

Variadic application, of the form

$$e \star e' \star e'' \star e''',$$

is explained below.

3. An *abstraction* of the form

$$binding \, . \, expression$$

denotes a *function*. The identifiers introduced are *variables*, with scope limited to the abstraction. Using f as an abbreviation with

$$f := v : X \wedge p \, . \, e,$$

the domain axiom is

$$d \in \mathscr{D}f \equiv d \in X \wedge p[_d^v$$

and the mapping axiom

$$d \in \mathscr{D}f \Rightarrow f\,d = e[_d^v.$$

Here, $e[_d^v$ is e with d substituted for v. Example: $n:\mathbb{Z}.2 \cdot n$.

4. *Tupling*, of the form e, e', e'' (any length n), denotes a function with domain $0..n-1$ and mapping illustrated by $(e, e', e'')\,0 = e$ and $(e, e', e'')\,1 = e'$ etc. The conditional expression $(p\,?\,e' \dagger e)$ is defined via tuples by

$$(p\,?\,e' \dagger e) = (e, e')\,p.$$

One can define shorthands ("macros") or sugaring in terms of the basic syntax, but very few suffice. Shorthands are d^e for $d \uparrow e$ (exponent) and d_e for $d \downarrow e$ (filtering, see below). Sugaring macros are

$$e \mid v:X \wedge p \qquad \text{for} \qquad v:X \wedge p.e,$$
$$v:X \mid p \qquad \text{for} \qquad v:X \wedge p.v,$$
$$v:=e \qquad \text{for} \qquad v:\iota\,e.$$

The last formula uses the *singleton set injector* ι with axiom

$$d \in \iota\,e \equiv d = e.$$

B.2.3   Style of Use

B.2.3.1   Functions

A function f is fully defined by its *domain* $\mathscr{D}f$ and its *mapping* (unique image for every domain element). Skipping a technicality, function equality can be expressed by

$$f = g \equiv \mathscr{D}f = \mathscr{D}g \wedge f © g,$$

where $f © g$ expresses that f and g are *compatible*, that is:

$$f © g \equiv \forall x:\mathscr{D}f \cap \mathscr{D}g.f\,x = g\,x.$$

Example: the *constant function definer* $\bullet$ with

$$X \bullet e = v:X.e$$

(v not free in e). It is near-trivial, but very useful.

Special instances: the *empty* function

$$\varepsilon := \emptyset^\bullet e$$

(any e; exercise) and the *one-point function definer* $\mapsto$ with

$$d \mapsto e = \imath\, d^\bullet e.$$

Predicates are $\mathbb{B}$-valued functions. We let $\mathbb{B} = \{0,1\}$; some may prefer $\mathbb{B} = \{\text{F},\text{T}\}$.

B.2.3.2 Operator Design

We show how to exploit the functional mathematics principle and (re)-synthesise common notations, issues that are not evident from mere syntax.

(a) *Elastic operators* originally are functionals designed to obviate common ad hoc abstractors such as $\Sigma_{i=m}^n$, $\forall v : X$, $\lim_{x \to a}$, but the idea leads to other designs as well.

   The *quantification* operators ($\forall$, $\exists$) are defined by

$$\forall P \equiv P = \mathscr{D}P^\bullet 1 \quad \text{and} \quad \exists P \equiv P \neq \mathscr{D}P^\bullet 0.$$

Observe synthesis of familiar expressions in

$$\forall P \equiv \forall x : \mathscr{D}P . P\,x \quad \text{and} \quad \forall x : \mathbb{R} . x^2 \geq 0$$

but also new forms as in

$$\forall (p,q) = p \wedge q \quad \text{and} \quad \exists (p,q) = p \vee q.$$

For every common infix operator $\star$ an *elastic extension* E is designed such that

$$x \star y = E(x,y).$$

Evident are $\bigcup$ and $\bigcap$ for $\cup$ and $\cap$, e.g.

$$e \in \bigcap S \equiv \forall x : \mathscr{D}S . e \in S\,x,$$

more interesting are Σ for $+$ (see below) and the following extensions for $=$ and $\neq$.

   The *constancy* predicate con and the *injectivity* predicate inj with

$$\mathsf{con} f \equiv \forall x : \mathscr{D}f . \forall y : \mathscr{D}f . f\,x = f\,y$$

$$\mathsf{inj} f \equiv \forall x : \mathscr{D}f . \forall y : \mathscr{D}f . f\,x = f\,y \Rightarrow x = y$$

follow the same design principle. Properties are

$$\mathsf{con}\,(d,e) \equiv d = e \quad\text{and}\quad \mathsf{inj}\,(d,e) \equiv d \neq e.$$

The *(function) range* operator $\mathscr{R}$ has axiom

$$e \in \mathscr{R}f \equiv \exists x : \mathscr{D}f . f\,x = e.$$

Using $\{-\}$ as a synonym for $\mathscr{R}$ synthesizes set notations such as

$$\{m : \mathbb{N} \mid m < n\} \quad\text{and}\quad \{2 \cdot n \mid n : \mathbb{Z}\}.$$

Since we never abuse "$\in$" for binding,

$$\{n : \mathbb{N} \mid n \in \mathbb{Z}\} \quad\text{and}\quad \{n \in \mathbb{N} \mid n : \mathbb{Z}\}$$

are unambiguous. Expressions like $\{e, e', e''\}$ also have their usual meaning. Rules are derived via $\exists$. We use $\mathscr{R}$ in defining the *function arrow* $\rightarrow$ by

$$f \in X \rightarrow Y \equiv \mathscr{D}f = X \wedge \mathscr{R}f \subseteq Y.$$

Variadic function application is alternating an infix operator with arguments. We *uniformly* take this as standing for the application of a matching elastic operator to the argument list. Examples:

$$p \wedge q \wedge r \equiv \forall(p,q,r) \quad\text{and}\quad e = e' = e'' \equiv \mathsf{con}\,(e,e',e'').$$

An example of a new opportunity is

$$e \neq e' \neq e'' \equiv \mathsf{inj}\,(e,e',e'').$$

Traditional ad hoc abstractors have a "range" attached to them, as in $\Sigma_{i=m}^{n}$. Elastic operators subsume this by the domain of the argument. This *domain modulation* principle obtains additional flexibility from he generic *function/set filtering* operator $\downarrow$ defined by

$$f_P = x : \mathscr{D}f \cap \mathscr{D}P \wedge P\,x . f\,x \quad\text{and}\quad X_P = \{x : X \cap \mathscr{D}P \mid P\,x\}.$$

(b) *Generic functionals* by design are applicable to *arbitrary* functions. Some extend existing functionals by lifting restrictions. For instance, function inversion f^{-} traditionally requires $\mathsf{inj}\,f$ and composition $f \circ g$ traditionally requires

$\mathscr{R}\,g \subseteq \mathscr{D}\,f$. We discard all restrictions on the arguments by defining the domain of the result such that its image definition is free of out-of-domain applications, e.g.

$$f \circ g = x : \mathscr{D}\,g \wedge g\,x \in \mathscr{D}\,f \,.\, f\,(g\,x).$$

Other generic functionals have no classical counterpart (Boute 2003). Examples are the *function override* ($\oslash$) and the *function merge* ($\cup$):

$$f \oslash g = x : \mathscr{D}\,f \cup \mathscr{D}\,g \,.\, x \in \mathscr{D}\,g\,?\,g\,x \,\dagger\, f\,x$$

$$f \cup g = x : \mathscr{D}\,f \cup \mathscr{D}\,g \wedge (x \in \mathscr{D}\,f \cap \mathscr{D}\,g \Rightarrow f\,x = g\,x) \,.\, (f \oslash g)\,x.$$

(c) *Two design examples*   For function types, a useful refinement of $\rightarrow$ is the Functional Cartesian Product $\times$ with

$$\times\,T = \{f : \mathscr{D}\,T \rightarrow {\textstyle\bigcup} T \mid \forall x : \mathscr{D}\,f \cap \mathscr{D}\,T \,.\, f\,x \in T\,x\}.$$

Some properties:

(i) $\times^{-}\,X\,x = \{f\,x \mid f : X\}$ for nonempty $X : \mathscr{R}\,\times$ and $x : \mathscr{D}\,(\times^{-}\,X)$;
(ii) $X \rightarrow Y = \times\,(X \bullet Y)$;
(iii) $X \times Y = \times\,(X, Y)$ where $X \times Y$ is the usual Cartesian product defined by

$$x, y \in X \times Y \equiv x \in X \wedge y \in Y.$$

Hence, we define variadic application of $\times$ by

$$X \times Y \times Z = \times\,(X, Y, Z)$$

etc. Often we write $X \ni v \rightarrow Y_v$ for $\times\,v : X \,.\, Y_v$ (known as a *dependent type*).

   A a generalized and formal definition of Σ is given via the generic function merge $\cup$. We define Σ recursively by:

The *empty rule* $\Sigma\,\varepsilon = 0$.
The *one-point rule* $\Sigma\,(e \mapsto c) = c$.
The *merge rule* $\Sigma\,(f \cup g) = \Sigma\,f + \Sigma\,g$

for any numeric c and any number-valued functions f and g with finite nonintersecting domains. The function domains may happen to be numeric. As expected, variadic application of $+$ is defined by $x + y + z = \Sigma\,(x, y, z)$.

   With —..— defined as in Pascal by $m..n = \{i : \mathbb{Z} \mid m \le i \le n\}$ for integer n and m, we formalize

$$\sum_{i=m}^{n} e \qquad \text{as standing for} \qquad \Sigma\,i : m..n\,.\,e.$$

This formally yields all rules, less the many pitfalls in common conventions.

B.2.3.3 Sequences, Sequence Types and Operators

Tuples, arrays, lists, here jointly called *sequences*, are the most ubiquitous aggregate data structures in discrete mathematics. Here we recast them in a functional mold. Letting

$$\mathbb{N}' := \mathbb{N} \cup \iota \infty,$$

we define the *block* operator

$$\square : \mathbb{N}' \to \mathscr{P} \mathbb{N} \quad \text{by} \quad \square n = \{m : \mathbb{N} \mid m < n\},$$

for instance, $\square 0 = \emptyset$ and $\square 2 = \mathbb{B}$ and $\square \infty = \mathbb{N}$.

A *sequence* is any function with domain $\square n$ for some $n : \mathbb{N}'$. The *length* operator # is defined by

$$\# x = n \equiv \mathscr{D} x = \square n$$

for any sequence x and $n : \mathbb{N}'$. The *empty sequence* is ε, and the *singleton sequence injector* τ is defined by $\tau e = 0 \mapsto e$.

An *array of length n over set A* is a function of type $\square n \to A$, written $A \uparrow n$ or A^n. Note that

$$A^n = \times (\square n \bullet A).$$

Finite lists over A are functions of type

$$\bigcup n : \mathbb{N}. A^n,$$

written A^*. Note that $A^\infty = \mathbb{N} \to A$ and $A^* \cap A^\infty = \emptyset$. We define A^ω as $A^* \cup A^\infty$.

Arbitrary tuple types are covered by $\times$. If T is a list of sets, $\times T \subseteq (\bigcup T)^{\# T}$.

An important operator on sequences is *concatenation* $\mathbin{+\!\!+}$. For any sequences x and y, we define $x \mathbin{+\!\!+} y$ to be the sequence with domain and mapping given respectively by

$$\#(x \mathbin{+\!\!+} y) = \#x + \#y \quad \text{and} \quad (x \mathbin{+\!\!+} y) i = (i < \#x) ? x i \mid y (i - \#x)$$

for all $i : \mathscr{D}(x \mathbin{+\!\!+} y)$. Useful properties are

identity: $x \mathbin{+\!\!+} \varepsilon = x = \varepsilon \mathbin{+\!\!+} x$ and associativity: $(x \mathbin{+\!\!+} y) \mathbin{+\!\!+} z = x \mathbin{+\!\!+} (y \mathbin{+\!\!+} z)$.

Derived operators are *prefixing* $\succ$ and *postfixing* $\prec$ with

$$e \succ x = \tau e \mathbin{+\!\!+} x \quad \text{and} \quad x \prec e = x \mathbin{+\!\!+} \tau e.$$

Alternatively, we can let $\succ$ be defined as the basic operator via its mapping and define $+\!\!+$ recursively by

$$\varepsilon +\!\!+ y = y \quad \text{and} \quad (a \succ x) +\!\!+ y = a \succ (x +\!\!+ y).$$

Induction for lists is expressed as follows: for any $P : A^* \to \mathbb{B}$

$$\forall P \equiv P\,\varepsilon \wedge \forall x : A^* . P\,x \Rightarrow \forall a : A . P(a \succ x).$$

B.2.3.4 Formal Calculation Rules

Readers with some experience in abstract algebra can infer the rules from the various definitions given; others can find an overview in (Boute 2005).

B.2.4 Introducing TLA$^+$ Via Funmath

Funmath is designed as a mathematical language without restrictions imposed by implementation concerns. As a result it is a good vehicle for introducing the syntax and semantics of TLA$^+$. The reader will notice how small the differences are. This is due to the fact that TLA$^+$ was also designed "close" to mathematics, and explains why it is preferable over program-like notations where restrictions inherited from implementation concerns are the rule. This closeness also allows describing the syntax by juxtaposition of equivalent expressions rather than using the full machinery of a formal grammar.

In the tables that follow, TLA$^+$ notations are organized in the order of their appearance on pages 268–269 of *Specifying Systems* by Leslie Lamport (2002a). The Funmath equivalents are expressed by means of functions and generic functionals defined above. The only function not listed there is the choice operator $\llbracket$, specified by

$$\mathscr{R} f \neq \mathbb{0} \Rightarrow \llbracket f \in \mathscr{R} f.$$

In Funmath terminology, an operator is an identifier for a function (a mathematical object, having a domain). In TLA$^+$, an operator is not a mathematical object and has no domain; pages 69–72 of *Functions versus Operators* by Leslie Lamport (2002a).

Tables B.1 (page 179) and B.2 (page 180) cover the basic mathematical expressions. Left columns pertain to TLA$^+$, right columns to Funmath. PF is the point-free variant, available in Funmath only, where it is the basis for the definition. Indices in $0..n-1$ and ellipsis (…) are just metanotation. In TLA$^+$ and Funmath, function applications (hence indexing) are not allowed in bindings, and ellipsis is illegal nonsense.

Table B.1   Basic mathematical TLA$^+$ expressions via Funmath equivalent, part 1

TLA$^+$	Funmath
Logic	
$\wedge$ $\vee$ $\neg$ $\Rightarrow$ $\equiv$	$\wedge$ $\vee$ $\neg$ $\Rightarrow$ $\equiv$
TRUE FALSE BOOLEAN	1 0 $\mathbb{B}$
$\forall x : p$ $\exists x : p$ $\forall x \in S : p$ $\exists x \in S : p$	$\forall x : S . p$ $\exists x : S . p$ PF: $\forall P$ $\exists P$
CHOOSE $x : p$ CHOOSE $x \in S : p$	$\llbracket x : S \mid p$ PF: $\llbracket f$
Sets	
$=$ $\neq$ $\in$ $\notin$ $\cap$ $\cup$ $\subseteq$ $\setminus$	$=$ $\neq$ $\in$ $\notin$ $\cap$ $\cup$ $\subseteq$ $\setminus$
$\{e_0, \ldots, e_{n-1}\}$	$\{e_0, \ldots, e_{n-1}\}$ PF: $\mathscr{R} s$
$\{x \in S : p\}$	$\{x : S \mid p\}$ PF: $\mathscr{R} f$
$\{e : x \in S\}$	$\{e \mid x : S\}$ PF: $\mathscr{R} f$
SUBSET S	$\mathscr{P} S$
UNION S (here S is a set of sets)	$\bigcup T$ (here T is a family of sets)
Functions	
$f[e]$	$f\,e$
DOMAIN f	$\mathscr{D} f$
$[x \in S \mapsto e]$	$x : S . e$
$[S \to T]$	$S \to T$
$[f$ EXCEPT $![d] = e]$	$f \oslash d \mapsto e$ PF: $f \oslash g$

The leftmost column comes from Leslie Lamport, *Specifying Systems* (Lamport 2002a, pp. 268–269)

We also recall that, in Funmath, x, y is always a tuple (notational consistency); hence,

TLA$^+$ binding	Funmath binding	
$x, y \in S$	$x, y : S^2$	
$\langle x, y \rangle \in T$	$x, y : T$	e.g. if $T = X \times Y$
$x \in X, y \in Y$	$x : X; y : Y$	

The main *action operators* and *temporal operators* are listed with their definition in Table B.3 (page 180) and are further explained in Sect. B.4.3.

B.3   Faithful Formalization of Informal Specifications

One can hardly overemphasize the importance of faithful formalization of informal statements or arguments. Indeed, many mistakes are made in "jumping ahead" by skipping [parts of] sentences or by translation into inadequate data structures. A formal specification that faithfully reflects the informal description is the best reference for later analysis or discussion to discover ambiguities, omissions, inconsistencies,

Table B.2 Basic mathematical TLA$^+$ expressions via Funmath equivalent, part 2

TLA$^+$	Funmath
Records	
$e.h$ as shorthand for $e[\text{"h"}]$	$e\,h$ (see note b. below)
$[h_0 \mapsto e_0, \ldots, h_{n-1} \mapsto e_{n-1}]$	$h_0 \mapsto e_0 \cup \cdots \cup h_{n-1} \mapsto e_{n-1}$
$[h_0 : S_0, \ldots, h_{n-1} : S_{n-1}]$	$\mathsf{Rcrd}\,(h_0 \mapsto S_0, \ldots, h_{n-1} \mapsto S_{n-1})$
$[r \text{ EXCEPT } !.h = e]$	$r \oslash (h \mapsto e)$

Notes on the Funmath column:

    a. Point-free versions of the last 3 lines: $\cup\,(h \mathbin{\widehat{\frown}} e)$ $\mathsf{Rcrd}\,(h \mathbin{\widehat{\frown}} S)$ $r \oslash s$

    b. The field name h is from an enumeration set; one can also use strings as in $e\text{"h"}$.

    c. $\mathsf{Rcrd}\,(h_0 \mapsto S_0, \ldots, h_{n-1} \mapsto S_{n-1}) = \bigtimes\,(h_0 \mapsto S_0 \cup \cdots \cup h_{n-1} \mapsto S_{n-1})$

Tuples	
$e[i]$	$e\,i$
$\langle e_0, \ldots, e_{n-1} \rangle$	$e_0, \ldots, e_{n-1}$ PF: e
$S_0 \times \ldots \times S_{n-1}$	$S_0 \times \cdots \times S_{n-1}$ PF: $\bigtimes S$
Strings and numbers	
$\text{"}c_0 \ldots c_{n-1}\text{"}$	$\text{"}c_0 \ldots c_{n-1}\text{"}$
STRING	$\mathbb{A}^*$ (for character set $\mathbb{A}$)
$d_{n-1} \ldots d_0 \quad d_{n-1} \ldots d_0 . d_{-1} \ldots d_{-m}$	$d_{n-1} \ldots d_0 \quad d_{n-1} \ldots d_0 . d_{-1} \ldots d_{-m}$
Miscellaneous	
IF p THEN e_1 else e_0	$p\,?\,e_1 \dagger e_0$
CASE $p_0 \to e_0 \,\Box\, \ldots \,\Box\, p_{n-1} \to e_{n-1}$	$\Box\,i : \Box\,n \wedge p_i . e_i$ PF: $\Box\,(f \downarrow P)$
CASE $p_0 \to \ldots \to e_{n-1} \,\Box\, \text{OTHER} \to d$	PF: $\exists P\,?\,(\Box\,(f \downarrow P)) \dagger d$
LET $c_0 \overset{\Delta}{=} e_0 \ldots c_{n-1} \overset{\Delta}{=} e_{n-1}$ IN d	$\mathbf{let}\ c_0 := e_0; \ldots; c_{n-1} := e_{n-1}\ \mathbf{in}\ d$
	$d\ \mathbf{where}\ c_0 := e_0; \ldots; c_{n-1} := e_{n-1}$

Note: in Funmath, a function CASE can be defined supporting any desired arrangement of conditions and expressions, e.g. CASE $(p_0 \mapsto e_0, \ldots, p_{n-1} \mapsto e_{n-1})$ (exercise).

The leftmost column comes from Leslie Lamport, *Specifying Systems* (Lamport 2002a, pp. 268–269)

Table B.3 Action and temporal operators of TLA$^+$ defined via Funmath

Action operators	
e'	$e' = e[^s_{s'}$
$[A]_e$	$[A]_e \equiv A \vee e' = e$
$\langle A \rangle_e$	$\langle A \rangle_e \equiv A \wedge e' \neq e$
ENABLED A	ENABLED $A \equiv \exists s' : \mathbf{S}.\ A$
UNCHANGED e	UNCHANGED $e \equiv e' = e$
$A \cdot B$	$A \cdot B \equiv \exists t : \mathbf{S}.\ A[^{s'}_t \wedge B[^s_t$
Temporal operators: primitive operators and patterns	
$\Box F$	$\beta \models \Box F \equiv \forall n : \mathbb{N}.\ \sigma^n \beta \models F$
$\Diamond F$	$\beta \models \Diamond F \equiv \exists n : \mathbb{N}.\ \sigma^n \beta \models F$
$\mathrm{WF}_e(A)$	$\mathrm{WF}_e(A) \equiv \Box(\Box(\text{ENABLED}\,\langle A \rangle_e) \Rightarrow \Diamond\langle A \rangle_e)$
$\mathrm{SF}_e(A)$	$\mathrm{SF}_e(A) \equiv \Box\Diamond\text{ENABLED}\,\langle A \rangle_e \Rightarrow \Box\Diamond\langle A \rangle_e$
$\forall x : F$	$\beta \models \forall x : F \equiv \forall \gamma : \mathbf{S}^\infty_{\models F} . (\natural\gamma)^\top_{\neq i} = (\natural\beta)^\top_{\neq i}$ where $i := \underset{\sim}{s}^- x$
$\exists x : F$	$\beta \models \exists x : F \equiv \exists \gamma : \mathbf{S}^\infty_{\models F} . (\natural\gamma)^\top_{\neq i} = (\natural\beta)^\top_{\neq i}$ where $i := \underset{\sim}{s}^- x$

The leftmost column comes from Leslie Lamport, *Specifying Systems* (Lamport 2002a, pp. 268–269)

inaccuracies and, most importantly, useful relationships (differences, equivalences) between formulations that are different in viewpoint or in the formal language used.

We start with simple examples of a non-technical nature, which best illustrate the point because no distracting technical side-issues are present.

B.3.1 Choice of Proper Data Abstractions

B.3.1.1 A Simple Example: The Coffee Bean Puzzle

At various places on the web, one finds the following puzzle, most likely originating from David Gries.

> Consider a coffee can which contains an unknown number of brown beans and an unknown number of white beans. Repeat the following process until exactly one bean remains. Select two beans from the can at random. If they are both the same color, throw them both out, but insert another brown bean. If they are different colours, throw the brown one away, but return the white one. What can you deduce about the colour of the last bean as a function of the initial number of black and white beans? Hint: find a useful invariant maintained by the process.

Here is the usual informal solution: the obvious invariant is the parity (i.e., being even or odd) of the number of white beans, so the one remaining bean is white iff the original number of white beans was odd. Clearly many steps are skipped – to appear "clever".

B.3.1.2 Faithful Formalization: The Inversion Criterion

How faithfully an informal text is formalized is to some extent a matter of taste. However, there is a reasonable working criterion: how well can the informal statement be reconstructed from the formalized one? For obvious reasons, we call this the *inversion criterion*.

The essence of faithful renderings is in using proper abstractions. Here, "proper abstraction" does *not* mean: away from the problem (to the contrary), but: away from the restrictions due to the specification language or its implementation.

Hence the crucial limiting factor is the expressiveness of the language. For instance, the statement of the bean problem indicates that non-deterministic choice must be supported. However, as we shall see, more is needed.

B.3.1.3 Formalization in the Guarded Command Language

Here is a possible rendering in Dijkstra's Guarded Command Language (Dijkstra 1997; Dijkstra and Scholten 1990).

```
┌──────────────────────── MODULE Nob ────────────────────────┐
│ EXTENDS Naturals, TLC                                          │
├────────────────────────────────────────────────────────────┤
```

VARIABLES w, b, ws $s \triangleq \langle w, b, ws \rangle$ CONSTANTS — Initially W white and B brown.

```
├────────────────────────────────────────────────────────────┤
```

$CBini \triangleq w = W \wedge b = B \wedge w \geq 0 \wedge b \geq 0 \wedge w + b > 0 \wedge ws = 0$

$CBinv \triangleq w\%2 = W\%2 \wedge w \geq 0 \wedge b \geq 0 \wedge w + b > 0$

$SlctAnyTwo \triangleq w + b \geq 2 \wedge ws' = \text{CHOOSE } k \in 0 \,..\, w : k \leq 2 \wedge 2 - k \leq b$

$AllSame \triangleq \neg(ws' = 1)$

$brown \triangleq 0 \quad white \triangleq 1$ Possible values for r in $Rplcby(r)$

$Rplcby(r) \triangleq w' = w - ws' + r \wedge b' = b - (2 - ws') + (1 - r)$

$JstOneLft \triangleq w + b = 1$

$CBfin \triangleq \text{UNCHANGED } s \wedge Print(\langle$ "Final number of white beans", $w\rangle, \text{TRUE})$

$CBnxt \triangleq \vee\, (SlctAnyTwo \wedge \text{IF } AllSame \text{ THEN } Rplcby(brown)$
$$\text{ELSE } Rplcby(white))$$
$\qquad\quad \vee\, (JstOneLft \wedge CBfin)$

$CB \triangleq CBini \wedge \Box[CBnxt]_s$

```
├────────────────────────────────────────────────────────────┤
```

THEOREM $CB \Rightarrow \Box CBinv$

```
└────────────────────────────────────────────────────────────┘
```

Fig. B.1 Numbers of beans

```
do w + b > 1 -> if w >= 2 -> w := w - 2 ; b := b + 1
                [] b >= 2 -> b := b - 1
                [] w >= 1 and b >= 1 -> b := b - 1    fi od
```

It deserves at least two relevant criticisms.

First, since this language supports non-determinism only via an if-statement, the random choice in the problem statement had to be "forced" into this shape.

Second, this rendering starts directly with numbers, which is coding-oriented and hence not a proper abstraction: the inversion criterion is far from being met.

Still, this program provides a nice and simple exercise for proving termination and invariants in a short classroom session or in an exam. For this purpose, we recommend the "checklist" given in (Gries and Schneider 1993) and proven in detail in (Boute 2006c).

B.3.1.4   Low-Level Formalization in TLA$^+$

Figure B.1 (*Nob* standing for "number of beans") formalizes the beans puzzle in TLA$^+$ (Lamport 2002a) in a low-level representation, that is: using numbers.

```
┌─────────────────────── MODULE Bob ───────────────────────┐
 EXTENDS Naturals, FiniteSets, Bags, TLC
 IsBag(B) ≜ B ∈ [DOMAIN B → {n ∈ Nat : n > 0}]  and some other auxiliaries
├──────────────────────────────────────────────────────────┤
 VARIABLES c, bs    s ≜ ⟨c, bs⟩  CONSTANT — Initial contents C
 white ≜ "wt"    brown ≜ "br"    Beans ≜ {white, brown}

 CBini ≜ c = C ∧ IsBagOf(c, Beans) ∧ c ≠ EmptyBag ∧ bs = EmptyBag
 CBinv ≜ ∧ CopiesIn(white, c)%2 = CopiesIn(white, C)%2
          ∧ IsBagOf(c, Beans) ∧ c ≠ EmptyBag

 SlctAnyTwo ≜ size(c) ≥ 2 ∧ bs' = CHOOSE p ∈ SubBag(c) : size(p) = 2
 AllSame ≜ Homog(bs')
 Rplcby(r) ≜ c' = c ⊖ bs' ⊕ SingBag(r)
 JstOneLft ≜ size(c) = 1

 CBfin ≜ UNCHANGED s ∧ Print(⟨"White beans", CopiesIn(white, c)⟩, TRUE)

 CBnxt ≜ ∨ (SlctAnyTwo ∧ IF AllSame THEN Rplcby(brown)
                                    ELSE Rplcby(white)
         ∨ (JstOneLft ∧ CBfin)
 CB ≜ CBini ∧ □[CBnxt]_s
├──────────────────────────────────────────────────────────┤
 THEOREM   CB ⇒ □CBinv
└──────────────────────────────────────────────────────────┘
```

Fig. B.2 Bag of beans

The central definition in this module is *CBnxt*. The example shows how to express procedural specifications by relations. Even before becoming proficient in formal calculation with relations, beginning students soon get some feeling for this different style of expression as compared to C++ or Java.

On the other hand, here is some style criticism. Note how the "meaningful identifiers" (*SlctAnyTwo* etc.) create the illusion of a faithful rendering (Dijkstra 1989), but the real test by the inversion criterion is after replacing these identifiers with their definitions a few lines earlier. All that remains then is some mumbo-jumbo with numbers, far removed from the problem statements. This is where data abstraction enters the picture, as illustrated in the following paragraph.

B.3.1.5   Formalization in TLA⁺ Using Bags

Figure B.2 (*Bob* standing for "bag of beans") uses the datastructure *Bags* to model both the contents of the can *c* and the bean sample *bs*.

The central definition *CBnxt* is unchanged w.r.t. Fig. B.1 mainly for educational reasons: emphasizing what is different, without possibly distracting other changes. However, replacing the "meaningful identifiers" with their definitions a few lines earlier now results in a procedure specification meeting the inversion criterion.

The only difference is that the generic data abstraction is called a "bag" whereas the problem statement talks about a "can". For instance,

$$bs' = \text{CHOOSE } p \in SubBag(c) : size(p) = 2$$

reads as: "[the sample] bs' comes from choosing any subbag from c of size 2".

B.3.2   Auxiliary Functions in Formal Specifications

In the preceding example modules, state transitions (from s to s') were very prominent, as is typical in TLA$^+$ specifications. However, the discussions showed the importance of the auxiliary data structures and operators, since these made the difference.

In experimenting with TLA$^+$, Funmath turns out to be a very good formalism for reasoning about function definitions. Here we illustrate two issues: styles of function definition for matching informal descriptions, and styles of definition for making definitions acceptable to TLC while preserving elegance and clarity. We start with the latter for the sake of continuity with the preceding section.

B.3.2.1   Redesigning TLA$^+$-Definitions for Acceptability to TLC

Whereas TLA$^+$ as a language imposes no restrictions beyond its syntax, TLC accepts only definitions that do not require exploring infinite structures.

For instance, Fig. B.3 shows the basic *Bags* module from (Lamport 2002a, p. 343), compressed by omitting some details that do not affect this discussion. A pleasant feature of TLA$^+$ is that it is using a mathematical rather than a programming notation. Hence, readers with a minimal general mathematical background (not specialist familiarity with some programming language) should be able to understand these definitions with a little bit of study.

Consider, however, the definition of *SubBag* in Fig. B.3, a function used in the *Bob* module (Fig. B.2). When model checking *BoB* using TLC, an error message indicated that the definition in Fig. B.3 leads TLC to exploring the natural numbers.

Here, we show how to develop a TLC-acceptable variant using the concepts of Funmath, in particular the Generalized Functional Cartesian Product (Sect. B.2), which is our "workhorse" for function typing (Boute 2003, 2005). We recall that, for any set-valued function T,

$$\times T = \{f : \mathscr{D}\,T \to \bigcup T \mid \forall x : \mathscr{D}\,T \cap \mathscr{D}f . f\,x \in T\,x\}. \tag{B.1}$$

The set of subbags of a given bag B is built as follows. The domain of any subbag of B is a subset of $\mathscr{D}\,B$, say S. A little reflection shows that the set of subbags of B

```
┌──────────────────────── MODULE SBagsC ─────────────────────────┐
│ LOCAL INSTANCE Naturals                                                          │
│ IsABag(B)  ≜  B ∈ [DOMAIN B → {n ∈ Nat : n > 0}]                                 │
│ SetToBag(S) ≜ [e ∈ S ↦ 1]                                                        │
│ BagIn(e, B) ≜ e ∈ DOMAIN B                                                       │
│ EmptyBag ≜ SetToBag({})                                                          │
│ CopiesIn(e, B) ≜ IF BagIn(e, B) THEN B[e] ELSE 0                                 │
│ B1 ⊕ B2  ≜  [e ∈ (DOMAIN B1)∪(DOMAIN B2) ↦ CopiesIn(e, B1) + CopiesIn(e, B2)]    │
│ B1 ⊖ B2  ≜  LET B ≜ [e ∈ DOMAIN B1 ↦ CopiesIn(e, B1) − CopiesIn(e, B2)]          │
│                IN   [e ∈ {d ∈ DOMAIN B : B[d] > 0} ↦ B[e]]                        │
│ LOCAL Sum(f) ≜ LET DSum[S ∈ SUBSET DOMAIN f] ≜                                    │
│                       LET elt ≜ CHOOSE e ∈ S : TRUE                              │
│                       IN   IF S = {} THEN 0 ELSE f[elt] + DSum[S \ {elt}]         │
│                    IN   DSum[DOMAIN f]                                            │
│ B1 ⊑ B2  ≜ (DOMAIN B1) ⊆ (DOMAIN B2) ∧ ∀e ∈ DOMAIN B1 : B1[e] ≤ B2[e]            │
│ SubBag(B) ≜ LET AllBagsOfSubset ≜                                                │
│                    UNION {[SB → {n ∈ Nat : n > 0}] : SB ∈ SUBSET DOMAIN B}       │
│                IN   {SB ∈ AllBagsOfSubset : ∀e ∈ DOMAIN SB : SB[e] ≤ B[e]}       │
│ BagCardinality(B) ≜ Sum(B)                                                        │
└──────────────────────────────────────────────────────────────────┘
```

Fig. B.3 The *Bags* module from *Specifying Systems* compressed

with domain S is $\times s:S.1..B\,s$ and hence the set of all subbags of B is the union of these sets as S ranges over all subsets of B, that is:

$$SubBag\,B = \bigcup S : \mathscr{P}(\mathscr{D}B). \times s:S.1..B\,s \qquad (B.2)$$

where $\mathscr{P}X$ is the powerset of set X. Expanding $\times s:S.1..B\,s$,

$$\times (s:S.1..B\,s)$$
$$= \langle \text{Def. } \times (B.1)\rangle \; \{b:S \to \bigcup(s:S.1..B\,s) \mid \forall s:S.b\,s \in 1..B\,s\}$$
$$= \langle \text{Calculations}\rangle \; \{b:S \to 1..nlub(B\,s \mid s:S) \mid b \sqsubseteq B\}$$

where *nlub* is the l.u.b. operator for $\mathbb{N}$ under $\leq$. We omit detail in "calculations". Substituting in (B.2) and translating the result into TLA$^+$ yields the following replacement for *SubBag* in Fig. B.3, bringing it within the capabilities of TLC.

B.3.2.2 Defining TLA$^+$ Functions at a More General Level

Once more we address faithful formalization of informal statements, taking an example relevant to temporal logic.

Here is the informal statement: *Given a sequence of symbols, replace successive appearances of the same symbol* (aptly called *stuttering* in the context of temporal logic) *by a single appearance of that symbol.* We call this *stuttering elimination.* It is informative trying to formalize this before reading any further.

In (Lamport 2002a) one finds the following formal specification: for any infinite sequence σ,

$$\natural\sigma \overset{\Delta}{=} \text{LET } f[n \in \textit{Nat}] \overset{\Delta}{=} \text{IF } n = 0 \text{ THEN } 0$$
$$\text{ELSE } \quad \text{IF } \sigma[n] = \sigma[n-1]$$
$$\text{THEN } f[n-1]$$
$$\text{ELSE } f[n-1]+1$$
$$S \overset{\Delta}{=} \{f[n] : n \in \textit{Nat}\}$$
$$\text{IN } \quad [n \in S \mapsto \sigma[\text{CHOOSE } i \in \textit{Nat} : f[i] = n]]$$

This definition is not evident and covers infinite sequences only.

A formula reflecting the intuitive simplicity of the specification is designed as follows (Boute 2006b). The basic idea is that any function can be decomposed as $f = \bigcup x : \mathscr{D}f \, . \, x \mapsto f\,x$. Hence for any finite or infinite sequence β we can write $\beta = \bigcup n : \mathscr{D}\beta \, . \, n \mapsto \beta\,n$. Re-indexing the domain points by skipping duplicates yields the definition

$$\natural\beta = \bigcup n : \mathscr{D}\beta \, . \, \Sigma\,(k : \Box\,n \, . \, \beta\,(k+1) \neq \beta\,k) \mapsto \beta\,n,$$

where we wrote $\natural$ to indicate that the argument can be any sequence (finite or infinite). One can prove (exercise) that $\natural$ equals $\natural$ restricted to infinite sequences. An even simpler formula is the following: for any sequence β,

$$\natural\beta \;=\; +\!\!\!+\; n : \mathscr{D}\beta \, . \, (n > 0 \wedge \beta\,(n-1) = \beta\,n)\,?\,\varepsilon \mid \tau\,(\beta\,n), \qquad \text{(B.3)}$$

where $+\!\!\!+$ is the concatenation operator, ε the empty sequence and $\tau\,e$ the sequence of length one containing just the element e.

B.4   Calculational Reasoning and Patterns in TLA$^+$

B.4.1   Capturing Temporal Logics by Temporal Calculi

The usual formulations of formal temporal logic (Manna and Pnueli 1991) follow the style of traditional formal logic oriented towards metamathematical issues. However, for introducing concepts of temporal logic to the practicing engineer, it is more appropriate casting them into a calculus with the smooth algebraic flavor so appreciated in classical mathematics, and presenting them as just another theory in the common framework.

A simple approach is viewing temporal operators as an algebra of functions on infinite sequences. Here, so-called *linear time* is assumed. So-called *branching time* yields a calculus where some rules are slightly different, but these are derived in the same way.

Starting from the model rather than pure axiomatics reflects the systems viewpoint, yet the usual axioms of temporal logic (Manna and Pnueli 1991) can still be made into textually identical theorems (Boute 1986), thus providing equal abstraction.

Different styles are possible, also depending on other desiderata. For instance, to support awareness when using a model checking tool, the calculus should be able to capture the tool's language. We illustrate later how to do this for TLA⁺. Yet, it is also conceptually helpful to start with a very elementary form of temporal calculus that captures the common concepts, independently of tools, re-using the results in calculi for specific tools.

B.4.2 A Functional Temporal Calculus (FTC)

B.4.2.1 Principle and Operator Definitions

A very basic temporal calculus is obtained by defining temporal operators as predicate transformers, the predicates of interest pertaining to system behaviors (infinite sequences of system states).

Formally, let $\mathbf{S}$ be the state space (instantaneous values). *Behaviors* (infinite sequences) are functions of type $\mathbb{N} \to \mathbf{S}$, also written $\mathbf{S}^\infty$. The predicates of interest are Boolean-valued functions over $\mathbf{S}^\infty$, hence of type $\mathsf{BP} := \mathbf{S}^\infty \to \mathbb{B}$.

Logical operators of FTC are just *pointwise extensions* of the usual propositional operators: for any infix operator $\star$ (say, $\wedge$, $\vee$, $\Rightarrow$, $\equiv$) and any β in $\mathbf{S}^\infty$,

$$(P \star Q)\beta \equiv P\beta \star Q\beta. \tag{B.4}$$

Operators of type $\mathbb{B}^2 \to \mathbb{B}$ (for $\wedge$, $\vee$, $\Rightarrow$, $\equiv$) or $\mathbb{B} \to \mathbb{B}$ (for $\neg$) are thereby overloaded to type $\mathsf{BP}^2 \to \mathsf{BP}$ or $\mathsf{BP} \to \mathsf{BP}$ ("predicate transformers"). Extension can be made explicit if desired (Boute 2003), but it is unambiguous here.

Temporal operators of FTC are predicate transformers of type $\mathsf{BP} \to \mathsf{BP}$, e.g.

$$\bigcirc \quad (\text{"next"}) \quad \text{defined by} \quad \bigcirc P\beta \equiv P(\sigma\beta) \tag{B.5}$$

$$\square \quad (\text{"henceforth"}) \quad \text{defined by} \quad \square P\beta \equiv \forall n : \mathbb{N}.\, P(\sigma^n\beta) \tag{B.6}$$

$$\lozenge \quad (\text{"eventually"}) \quad \text{defined by} \quad \lozenge P\beta \equiv \exists n : \mathbb{N}.\, P(\sigma^n\beta) \tag{B.7}$$

By convention in functional formalisms, $f\, x\, y$ is read $(f\, x)\, y$, so $\square P\beta = (\square P)\beta$. Also, σ is the *shift* operator defined on any sequence s by $\sigma\, s\, m = s\,(m+1)$. Informally: σ drops the first symbol, e.g. $\sigma\,(a, b, c, d) = b, c, d$. The n-th power of a function is n-fold composition: $f^0\, x = x$ and $f^{n+1}\, x = f\,(f^n\, x)$ inductively.

By these definitions, FTC reduces temporal reasoning to predicate calculus.

Convention As in functional predicate calculus, $\forall P$ expresses that predicate P is satisfied by all elements in its domain (Boute 2005). However, to highlight analogy with expressions of the form $\vdash \varphi$ in typical temporal logics (Manna and Pnueli 1991), we define $\vdash$ for predicates P in BP by $\vdash P \equiv \forall P$ (in pointwise form, $\vdash P \equiv \forall \beta : \mathbf{S}^\infty . P\beta$).

Aside: in formal logic, $\vdash$ is usually a metasymbol for "theoremhood". Adopting $\vdash$ within the language, as done here, adds flexibility for elucidating analogies and paradigm shifts. Also, to the "working mathematician" *provability* and *validity* are tantamount. Lamport (2002a, p. 92) simply states *"A temporal theorem is a temporal formula that is satisfied by all behaviors"*.

B.4.2.2   Illustration: Deriving Point-Free Theorems in FTC

By *point-free* style we mean avoiding references to domain elements of functions (Boute 2003). Here the domain elements are the behaviours, typically referenced by a variable β (of type $\mathbf{S}^\infty$).

The point-free style allows writing formulas looking formally identical to the meta-theorems and axioms of typical temporal logics (Manna and Pnueli 1991).

The first stage in building this collection is deriving formulas by predicate calculus and getting rid of the variable β along the way. The second stage is using only point-free formulas already obtained, as a matter of style.

To convey the flavour, here are a few first-stage examples, selected assuming only little knowledge of predicate calculus, yet each yielding some interesting insight.

Example A   Showing $\vdash (\Box P) \equiv \vdash P$.

$$\vdash (\Box P) \;\equiv\; \langle \text{Definition} \vdash \rangle \;\; \forall \beta : \mathbf{S}^\infty . \Box P \beta$$

$$\equiv\; \langle \text{Definition} \,\Box \rangle \;\; \forall \beta : \mathbf{S}^\infty . \forall n : \mathbb{N} . P(\sigma^n \beta) \qquad (*)$$

$$(*) \;\Rightarrow\; \langle \text{Inst. } n := 0 \rangle \;\; \forall \beta : \mathbf{S}^\infty . P(\sigma^0 \beta)$$

$$\equiv\; \langle f^0 x = x \rangle \;\; \forall \beta : \mathbf{S}^\infty . P\beta$$

$$\equiv\; \langle \text{Definition} \vdash \rangle \;\; \vdash P$$

$$(*) \;\Leftarrow\; \langle \text{Inst. } \beta := \sigma^n \beta \rangle \;\; \forall \beta : \mathbf{S}^\infty . \forall n : \mathbb{N} . \forall \beta : \mathbf{S}^\infty . P\beta$$

$$\equiv\; \langle \text{Definition } \vdash \rangle \;\; \forall \beta : \mathbf{S}^\infty . \forall n : \mathbb{N} . \vdash P$$

$$\Leftarrow\; \langle \text{Const. pred.} \rangle \;\; \vdash P \;\; .$$

In better taste than such a 'ping-pong' argument, is an equational proof, but this requires a little more experience (exercise).

Example B, "temporal instantiation": $\Box P \Rightarrow P$.

A basic proof would consist in showing $\vdash (\Box P \Rightarrow P)$ to be equivalent to 1 (exercise). Observe that $P \Rightarrow \Box P$ is not a theorem (try $\beta\, n = n$ and $P\beta \equiv \beta\, 0 = 0$).

Remark Whereas such a takes 6 steps, a theorem of the form $\vdash Q$ can be proven more compactly by proving $Q\beta$ for arbitrary β. For the $\Box P \Rightarrow P$ example,

$$\Box P\beta \equiv \langle \text{Definition } \Box \rangle \; \forall n : \mathbb{N}. P(\sigma^n \beta)$$
$$\Rightarrow \langle \text{Inst. } n := 0 \rangle \; P(\sigma^0)$$
$$\equiv \langle \text{Definition } f^n \rangle \; P\beta,$$

which shows $\Box P\beta \Rightarrow P\beta$ and hence, by point-wise extension (B.4), $(\Box P \Rightarrow P)\beta$.

Example C: induction This example involves a nice generalization of the *weak induction principle* over natural numbers (WIN). A typical form of WIN is the following: for any predicate $Q : \mathbb{N} \to \mathbb{B}$,

$$\forall (n:\mathbb{N}. Q\, n \Rightarrow Q(n+1)) \Rightarrow (Q0 \Rightarrow \forall m:\mathbb{N}. Q\, m). \tag{B.8}$$

The converse does not hold (try $Q\, n \equiv n = 1$). However, calculation (exercise) also yields a *strengthened weak induction principle* over $\mathbb{N}$ (SWIN):

$$\forall (n:\mathbb{N}. Q\, n \Rightarrow Q(n+1)) \equiv \forall (n:\mathbb{N}. Q\, n \Rightarrow \forall m:\mathbb{N}. Q(n+m)), \tag{B.9}$$

from which WIN (B.8) is easily recovered by instantiating the r.h.s. with $n := 0$.

SWIN also yields a temporal counterpart in FTC by showing calculationally that $\Box(P \Rightarrow \bigcirc P)\beta \equiv \Box(P \Rightarrow \Box P)\beta$ (exercise). Hence $\Box(P \Rightarrow \bigcirc P) = \Box(P \Rightarrow \Box P)$ as function equality or, in temporal theorem style, the *strengthened (weak) temporal induction* principle (STI).

$$\vdash (\Box(P \Rightarrow \bigcirc P) \equiv \Box(P \Rightarrow \Box P)). \tag{B.10}$$

The two sides are a first instance of what can be considered a *pattern*, i.e., a composite temporal formula with some general nontrivial useful property. Here is another pattern.

Example D, "infinitely often" Writing $\Box\Diamond\varphi$ is typical in specifications to express that a formula φ is satisfied "infinitely often". Note that this is a model-centric statement. More importantly, it is nearly always given without justification, perhaps assuming the intuitive interpretation "no matter how often φ has already happened, it will happen again". So a formal proof is all the more revealing.

In mathematics, the usual characterization of finiteness is by correspondence to the set of the first n natural numbers for some n, or a subset thereof: for any predicate Q,

$$\mathsf{Fin}\, Q \equiv \exists n:\mathbb{N}. \exists f:\mathbb{N}_{<n} \to \mathscr{D}\, Q. (\mathscr{D}\, Q)_Q \subseteq \mathscr{R} f. \tag{B.11}$$

Infiniteness is just the negation of finiteness, so we define $\exists_\infty Q \equiv \neg(\mathsf{Fin}\, Q)$.

Specializing to predicates on natural numbers ($\mathscr{D} Q = \mathbb{N}$) allows showing that $\exists_\infty Q \equiv \forall n : \mathbb{N} . \exists m : \mathbb{N} . Q(m+n)$. The proof is quite instructive (exercise).

It simply follows that, for any temporal predicate P in BP and any β in $\mathbf{S}^\infty$,

$$\Box(\Diamond P)\beta \equiv \exists_\infty n : \mathbb{N} . P(\sigma^n \beta). \tag{B.12}$$

This formally proves that $\Box(\Diamond P)$ is indeed equivalent to P being satisfied "infinitely often" according to the common mathematical characterization.

Example E, distributivity(-like) properties   An important batch is

Distribut. $\Box/\wedge$: $\Box(P \wedge Q) \equiv \Box P \wedge \Box Q$	Dual: $\Diamond(P \vee Q) \equiv \Diamond P \vee \Diamond Q$
Dispatch. $\Diamond/\wedge$: $\Diamond(P \wedge Q) \Rightarrow \Diamond P \wedge \Diamond Q$	Dual: $\Box(P \vee Q) \Leftarrow \Box P \vee \Box Q$
Equal predic.: $\Box(P \equiv Q) \Rightarrow (\Box P \equiv \Box Q)$	Also: $\Box(P \equiv Q) \Rightarrow (\Diamond P \equiv \Diamond Q)$
Weaker predic.: $\Box(P \Rightarrow Q) \Rightarrow \Box P \Rightarrow \Box Q$	Also: $\Box(P \Rightarrow Q) \Rightarrow \Diamond P \Rightarrow \Diamond Q$

These are similar to certain properties in functional predicate calculus (Boute 2005). A noteworthy addition is $\exists_\infty(P \vee Q) \equiv \exists_\infty P \vee \exists_\infty Q$ for general predicates P and Q satisfying $\mathscr{D} P = \mathscr{D} Q$. For temporal predicates P and Q in BP, this yields $\Box(\Diamond(P \vee Q)) \equiv \Box(\Diamond P) \vee \Box(\Diamond Q)$; dual: $\Diamond(\Box(P \wedge Q)) \equiv \Diamond(\Box P) \wedge \Diamond(\Box Q)$.

We conclude this subsection with two important observations.

(a) FTC is entirely formulated within functional predicate calculus, without a separate temporal logic language. The operators are predicate transformers.

(b) FTC captures the essence of temporal logic, and can thereby serve as an archetype for studying existing temporal logics, an issue addressed next. As an example, we chose Lamport's Temporal Logic of Actions (Lamport 2002a) or TLA$^+$, and show how the aforementioned approach captures it as TCA. Since (Lamport 2002a) is readily available on the web, no detailed account of TLA$^+$ is necessary here.

B.4.3 Defining the Temporal Calculus of Actions (TCA)

B.4.3.1 Types

When dealing with functions, it is convenient to have their types at hand.

Although TLA$^+$ is untyped, types for the variables follow from an initial state and a next state specification. Moreover, as in (Lamport 2002a), a well-structured specification is documented by a *type invariant* stating types explicitly. Hence in the sequel we pretend that variables have been declared with a type.

Let $T : I \to \mathscr{T}$ be the family of types for the variables in the specification. The index set I is for bookkeeping, and can be tuned to the desired style. Then the state space is $\mathbf{S} := \bigtimes T$, a Cartesian product. Behaviours have type $\mathbf{S}^\infty$.

We assume that basic arithmetic, relational and logical operators are available on these types (language-dependent, but details not needed), and categorize expressions as follows.

$\mathscr{E}$	state expressions	$\mathscr{B}$ state propositions
$\mathscr{E}'$	transition expressions	$\mathscr{A}$ transition propositions, called *"actions"*
$\mathscr{X}$	temporal expressions	$\mathscr{F}$ temporal propositions (*"temporal formulas"*)

In state expressions, state variables occur unprimed, in transition expressions they can also be primed, and in temporal expressions temporal operators (see below) can occur, so $\mathscr{E} \subset \mathscr{E}' \subset \mathscr{X}$. Propositions are just boolean-valued expressions, so $\mathscr{B} \subset \mathscr{E}$, $\mathscr{A} \subset \mathscr{E}'$, $\mathscr{F} \subset \mathscr{X}$.

B.4.3.2 Conventions

Substituting an expression d for variable v in expression e is written $e[v := d]$ as in (Gries and Schneider 1993), or as $e[\begin{smallmatrix} v \\ d \end{smallmatrix}]$ (saving scarce horizontal space). For multiple substitution, d and v can be tuples (of the same length), for instance, $(y + x)[\begin{smallmatrix} x,y \\ z\cdot y, a\cdot x \end{smallmatrix}] = a \cdot x + z \cdot y$.

As in (Boute 2006c), s is a syntactic shorthand standing in all bindings and mathematical expressions for the tuple formed by all state variable names in some fixed order (e.g. as declared). The tuple of the names of the variables as syntactic elements is written $\underset{\sim}{s}$. The set of variables is then $\mathscr{V} := \mathscr{R} \underset{\sim}{s}$ and we let $I := \mathscr{D} \underset{\sim}{s}$. So, for n state variables, the state space $\mathbf{S}$ is a set of n-tuples. Furthermore, for any expression e, we write e' for $e[\begin{smallmatrix} s \\ s' \end{smallmatrix}]$, noting also that $s' = s[\begin{smallmatrix} s \\ s' \end{smallmatrix}]$.

Example: given the declaration VAR $num : \mathbb{Z}; cond : \mathbb{B}$, then $\mathbf{S} := \mathbb{Z} \times \mathbb{B}$ and s literally stands for $num, cond$, e.g. $\forall s : \mathbf{S}. p$ stands for $\forall (num, cond) : \mathbf{S}. p$ (parentheses optional).

B.4.3.3 Operators

In the tables, the leftmost columns describe the syntax via the syntactic categories (unusual but self-explanatory). The rightmost columns give translation into common notation.

a. Action operators

$$
\begin{array}{|l|l|}
\hline
\underline{}\cdot\underline{} : \mathscr{A} \times \mathscr{A} \to \mathscr{A} & a \cdot b \equiv \exists t : \mathbf{S}. a[\begin{smallmatrix} s' \\ t \end{smallmatrix}] \wedge b[\begin{smallmatrix} s \\ t \end{smallmatrix}] \\
[\underline{}]\underline{} : \mathscr{A} \times \mathscr{E} \to \mathscr{A} & [a]_e \equiv a \vee e = e' \\
\langle\underline{}\rangle\underline{} : \mathscr{A} \times \mathscr{E} \to \mathscr{A} & \langle a \rangle_e \equiv a \wedge e \neq e' \\
\text{UNCHANGED} : \mathscr{E} \to \mathscr{A} & \text{UNCHANGED}\, e \equiv e = e' \\
\text{ENABLED} : \mathscr{A} \to \mathscr{B} & \text{ENABLED}\, a \equiv \exists s' : \mathbf{S}. a \\
\hline
\end{array}
\qquad \text{(B.13)}
$$

b. Temporal operators are characterized using the endosemantic function $\models$, defined recursively on the structure of expressions. In view of $\mathcal{E} \subset \mathcal{E}' \subset \mathcal{X}$, the recursion basis are the state and transition expressions e in $\mathcal{E}$ and $\mathcal{E}'$, for which

$$\beta \models e \;=\; e\genfrac{[}{}{0pt}{}{s,s'}{\beta 0,\beta 1}.$$

Since state expressions contain no primed state variables, $e\genfrac{[}{}{0pt}{}{s,s'}{\beta 0,\beta 1} = e\genfrac{[}{}{0pt}{}{s}{\beta 0}$.

For temporal expressions (in $\mathcal{X}$) and formulas (in $\mathcal{F}$).

$$
\begin{array}{|ll|}
\hline
\bigcirc : \mathcal{X} \to \mathcal{X} & \beta \models \bigcirc e \;=\; \sigma\beta \models e \quad (\bigcirc \text{ is not part of TLA}^+) \\
\square : \mathcal{F} \to \mathcal{F} & \beta \models \square\varphi \;\equiv\; \forall n : \mathbb{N}.\, \sigma^n \beta \models \varphi \\
\Diamond : \mathcal{F} \to \mathcal{F} & \beta \models \Diamond\varphi \;\equiv\; \exists n : \mathbb{N}.\, \sigma^n \beta \models \varphi \\
\forall\_ : \mathcal{V} \to \mathcal{F} \to \mathcal{F} & \beta \models \forall_v \varphi \;\equiv\; \forall \gamma : \mathbf{S}^\infty_{\models \varphi}.\, (\natural\gamma)^\mathsf{T}_{\neq i} = (\natural\beta)^\mathsf{T}_{\neq i} \quad i = \underset{\sim}{s}{}^- v \\
\exists\_ : \mathcal{V} \to \mathcal{F} \to \mathcal{F} & \beta \models \exists_v \varphi \;\equiv\; \exists \gamma : \mathbf{S}^\infty_{\models \varphi}.\, (\natural\gamma)^\mathsf{T}_{\neq i} = (\natural\beta)^\mathsf{T}_{\neq i} \quad i = \underset{\sim}{s}{}^- v \\
\hline
\end{array}
\tag{B.14}
$$

The temporal quantifiers $\forall$ and $\exists$ are given for completeness only and may be skipped. The rather terse notation uses generic transposition from (Boute 2003): $f^\mathsf{T} y x = f\, x\, y$), and the *compacting operator* $\natural$ removes successive duplicates (stuttering); see Sect. B.3.2.

Boolean combinations of temporal formulas are defined by pointwise distributivity:

$$
\begin{array}{|ll|}
\hline
\beta \models \neg\varphi \equiv \neg(\beta \models \varphi) & \beta \models \forall\,(x : X.\,\varphi) \equiv \forall x : X.\,\beta \models \varphi \\
\beta \models (\varphi \star \psi) \equiv \beta \models \varphi \star \beta \models \psi & \beta \models \exists\,(x : X.\,\varphi) \equiv \exists x : X.\,\beta \models \varphi \\
\hline
\end{array}
\tag{B.15}
$$

Here, $\star$ is any infix logical operator in $\{\Rightarrow, \equiv, \not\equiv, \oplus, \wedge, \vee\}$.

At the left-hand sides of the equivalences, $\star$, $\neg$, $\forall$, $\exists$ are TCA/TLA$^+$ operators, and at the right-hand side they are the "normal" logical operators. Risk of confusion is minor, since the calculation rules will be fully analogous. Because temporal formulas appear only syntactically, we can adopt the syntax of the target language, e.g. for optional parentheses, $\square\Diamond\varphi$ stands for $\square(\Diamond\,\varphi)$ etc.

B.4.4 Calculational Reasoning in TCA/TLA$^+$

B.4.4.1 Introduction

For any φ, the partial application $\models\varphi$ is a predicate of type BP. This differs from FTC predicates only in using postfix notation (β stands before $\models\varphi$ in $\beta \models \varphi$, but after P in $P\beta$). Up to this lexical detail, all calculation rules are inherited from FTC.

The $\vdash$ operator from is adapted (or overloaded) to temporal formulas φ by

$$\vdash \varphi \;\equiv\; \forall \beta : \mathbf{S}^\infty . \beta \models \varphi. \tag{B.16}$$

Hence, $\vdash \varphi$ expresses the fact that φ is a (temporal) theorem in Lamport's sense (Lamport 2002a).

A *(temporal) tautology* is a (temporal) theorem containing only arbitrary formulas, which can be instantiated by specific ones as desired.

"Proving φ" then means "proving $\vdash \varphi$" but, as for FTC, expanding $\vdash \varphi$ in pointwise form according to (B.16) is necessary only in proving the basic theorems.

B.4.4.2 A Calculational Style for TCA/TLA$^+$

First, all calculations from FTC are inherited. It suffices replacing $P\beta$ by $\beta \models \varphi$ and, when desired, removing optional parentheses, e.g. duality $\Diamond P \equiv \neg (\Box(\neg P))$ becomes $\Diamond \varphi \equiv \neg \Box \neg \varphi$.

Basic tautologies are named correspondingly, again (as in FTC) borrowing the terminology from similar rules in general predicate calculus (Boute 2005), for instance

$$\Box(\varphi \wedge \psi) \equiv \Box \varphi \wedge \Box \psi \;(\text{Dist.}\,\Box/\wedge) \quad \Box(\varphi \vee \psi) \Leftarrow \Box \varphi \vee \Box \psi \;(\text{Coll.}\,\Box/\vee)$$
$$\Diamond(\varphi \vee \psi) \equiv \Diamond \varphi \vee \Diamond \psi \;(\text{Dist.}\,\Diamond/\vee) \quad \Diamond(\varphi \wedge \psi) \Rightarrow \Diamond \varphi \wedge \Diamond \psi \;(\text{Disp.}\,\Diamond/\wedge) \tag{B.17}$$

We also recall the extra equational distributivity rules due to the underlying model (behaviours) and based on properties of the natural numbers, e.g.

$$\Box\Diamond(\varphi \vee \psi) \equiv \Box\Diamond \varphi \vee \Box\Diamond \psi \quad (\text{Dist.}\,\Box\Diamond/\vee)$$
$$\Diamond\Box(\varphi \wedge \psi) \equiv \Diamond\Box \varphi \wedge \Diamond\Box \psi \quad (\text{Dist.}\,\Diamond\Box/\wedge). \tag{B.18}$$

Rules for "equal/weaker predicates", e.g. $\forall (P \Rrightarrow Q) \Rightarrow \forall P \Rightarrow \forall Q$ (WKP\\$\forall$) from general predicate calculus (Boute 2005) and rule $\Box(P \Rightarrow Q) \Rightarrow \Box P \Rightarrow \Box Q$ (WKP\\$\Box$) from FTC are renamed with "formula", as in $\Box(\varphi \Rightarrow \psi) \Rightarrow \Box \varphi \Rightarrow \Box \psi$ (WKF\\$\Box$).

Even when calculating directly with $\beta \models \varphi$, the remark after *Example B* in Sect. B.4.2 shows how to omit repetitive parts, such as the prelude "We calculate, for arbitrary $\beta : \mathbf{S}^\infty$," and the postlude "Hence $\beta \models \varphi \star \beta \models \psi$ which yields $\vdash (\varphi \star \psi)$", where $\star$ is implication or equivalence as in the calculation chain.

Finally, we establish a calculational style *within* TCA/TLA$^+$ as follows. Thus far, all steps in all derivations were linked by propositional equivalences and implications, and β appeared explicitly. However, after deriving the $\Box/\Diamond$-related tautologies, further calculations typically contain (only) steps of the form

$$\beta \models \varphi \;\Rightarrow\; \langle \text{Justification for } \beta \models \varphi \Rightarrow \beta \models \psi \rangle \quad \beta \models \psi$$
$$\beta \models \varphi \;\equiv\; \langle \text{Justification for } \beta \models \varphi \equiv \beta \models \psi \rangle \quad \beta \models \psi.$$

The justifications can be temporal tautologies of the form $\varphi \Rightarrow \psi$ or $\varphi \equiv \psi$, since these can be instantiated for $\beta : S^\infty$ using (B.16). There are more tautologies than just $\Box/\Diamond$-related ones. Every rule from propositional calculus yields a temporal tautology by substituting $\beta \models \varphi$, $\beta \models \psi$ etc. for p, q etc., distributivity (B.15) for every operator to bring $\beta \models$ in front, and generalization for $\forall$.

As $\beta \models$ appears in every line in the same position, we omit it as a matter of convention, linking the steps by temporal equivalences and implications.

All these observations are illustrated in the following calculation, yielding the interesting modus ponens-like property $\Diamond\Box\varphi \wedge \Diamond\Box(\varphi \Rightarrow \psi) \Rightarrow \Diamond\Box\psi$.

$$\Diamond\Box\varphi \wedge \Diamond\Box(\varphi \Rightarrow \psi) \equiv \langle\text{Dist. } \Diamond\Box/\wedge\rangle \; \Diamond\Box(\varphi \wedge (\varphi \Rightarrow \psi))$$

$$\equiv \langle\text{MP equiv.}\rangle \; \Diamond\Box(\varphi \wedge \psi)$$

$$\equiv \langle\text{Dist. } \Diamond\Box/\wedge\rangle \; \Diamond\Box\varphi \wedge \Diamond\Box\psi$$

$$\Rightarrow \langle\text{Weakening}\rangle \; \Diamond\Box\psi.$$

Rule $\langle\text{MP equiv.}\rangle$ is "Modus Ponens as an equivalence": $\varphi \wedge (\varphi \Rightarrow \psi) \equiv \varphi \wedge \psi$. The redundancy is to obtain $\Diamond\Box\varphi \wedge \Diamond\Box(\varphi \Rightarrow \psi) \equiv \Diamond\Box\varphi \wedge \Diamond\Box\psi$ in passing.

Calculation is now fully *within* TCA. When possible, we use this style as it reduces writing, makes patterns conspicuous, and raises the abstraction level.

B.4.5   Applications to Patterns in TLA$^+$

This section is an extended chain of examples about patterns related to liveness and fairness. The patterns are taken from Chap. 8 in *Specifying Systems* (Lamport 2002a), which is readily available on the web. We show how TCA yields significantly simpler proofs and how the theorems themselves are *discovered* by calculation, sometimes even in a stronger form. Formulas are labeled as in the cited reference.

B.4.5.1   Weak Fairness

From (Lamport 2002a, pp. 97–98), we quote the following equivalent patterns for "weak fairness", denoted $\text{WF}_v(A)$. The motivation is discussed in the cited reference.

$$\Box(\Box\text{ENABLED}\,\langle A\rangle_v \Rightarrow \Diamond\langle A\rangle_v) \quad \text{(8.7) in (Lamport 2002a)} \qquad \text{(B.19)}$$

$$\Box\Diamond(\neg\,(\text{ENABLED}\,\langle A\rangle_v) \vee \Box\Diamond\langle A\rangle_v \qquad \text{(8.8) in (Lamport 2002a)} \qquad \text{(B.20)}$$

$$\Diamond\Box(\text{ENABLED}\,\langle A\rangle_v) \Rightarrow \Box\Diamond\langle A\rangle_v \qquad \text{(8.8) in (Lamport 2002a)} \qquad \text{(B.21)}$$

These will be the basis for the following calculational TCA/TLA$^+$-derivations.

B.4.5.2 Application Example A

The motivation of (8.7) in (Lamport 2002a, p. 97) went via the intermediate form $\Box(\text{ENABLED}\,\langle A\rangle_v \Rightarrow \Diamond\langle A\rangle_v)$, later giving rise (Lamport 2002a, p. 99) to the question under which condition this form is equivalent to (8.7).

The answer in (Lamport 2002a, p. 99) is given in the form of a theorem:

$$\Box(E \Rightarrow \Box E \vee \Diamond A) \;\Rightarrow\; (\Box(E \Rightarrow \Diamond A) \equiv \Box(\Box E \Rightarrow \Diamond A)).$$

$$(8.11)\text{ in (Lamport 2002a)} \tag{B.22}$$

Formula (8.11) was designated as "complicated" and unfavorable to a proof by calculation, and a classical proof taking about one page was given.

Here follows a calculational *derivation*, which differs from a proof in the sense that the desired condition is *discovered* without knowing it in advance.

$$
\begin{aligned}
\Box(E \Rightarrow \Diamond A) \equiv \Box(\Box E \Rightarrow \Diamond A) \;&\Leftarrow\; \langle\text{Equal form.}\backslash\Box\rangle\; \Box(E \Rightarrow \Diamond A \equiv \Box E \Rightarrow \Diamond A) \\
&\equiv\; \langle\text{RSDist. }\Rightarrow/\equiv\rangle\; \Box(\neg(E \equiv \Box E) \Rightarrow \Diamond A) \\
&\equiv\; \langle\text{Inst. }\Box\varphi \Rightarrow \varphi\rangle\; \Box(\neg(E \Rightarrow \Box E) \Rightarrow \Diamond A) \\
&\equiv\; \langle\text{From }\Rightarrow\text{ to }\vee\rangle\; \Box((E \Rightarrow \Box E) \vee \Diamond A) \\
&\equiv\; \langle\text{From }\Rightarrow\text{ to }\vee\rangle\; \Box(\neg E \vee \Box E \vee \Diamond A) \\
&\equiv\; \langle\text{From }\vee\text{ to }\Rightarrow\rangle\; \Box(E \Rightarrow \Box E \vee \Diamond A)
\end{aligned}
$$

The "Right SemiDistributivity $\Rightarrow/\equiv$" rule is $\quad \neg(p \equiv q) \Rightarrow r \equiv p \Rightarrow r \equiv q \Rightarrow r$.

B.4.5.3 Application Example B

In (Lamport 2002a, p. 101 ff.), the question is asked when separate fairness conditions can be combined in a single one, more specifically, *When can $WF_v(A) \wedge WF_v(B)$ be written as $WF_v(A \vee B)$?*

The answer in (Lamport 2002a, p. 102) is given in the form of a theorem:

$$DR1 \wedge DR2 \;\Rightarrow\; (WF_v(A) \wedge WF_v(B) \equiv WF_v(A \vee B)) \text{ (8.20) in (Lamport 2002a)}$$

where

$$
\begin{aligned}
DR1 &\stackrel{\Delta}{=} \Box(\text{ENABLED}\,\langle A\rangle_v \Rightarrow \Box\neg\text{ENABLED}\langle B\rangle_v \vee \Diamond\langle A\rangle_v) \\
DR2 &\stackrel{\Delta}{=} \Box(\text{ENABLED}\,\langle B\rangle_v \Rightarrow \Box\neg\text{ENABLED}\langle A\rangle_v \vee \Diamond\langle B\rangle_v).
\end{aligned}
$$

The classical proof in (Lamport 2002a, p. 102 ff.) takes two and a half pages. It uses contradiction, which also requires knowing the result. Here, we proceed calculationally, and by discovery, which happens to yield a stronger result along the way.

To avoid clutter in formulas and calculations, we introduce W defined by the following equivalent expressions for W A, from which to choose as convenient.

$$\Box(\Box \text{E}\, A \Rightarrow \Diamond A) \quad \Box\Diamond\neg\text{E}\, A \vee \Box\Diamond A \quad \Diamond\Box\text{E}\, A \Rightarrow \Box\Diamond A \qquad (B.23)$$

Obviously $WF_v(A) \equiv \text{W}\langle A\rangle_v$. Note also that E and $\langle\ \rangle_v$ distribute over $\vee$.

The central question is when $WF_v(A \vee B)$ captures $WF_v(A) \wedge WF_v(B)$. Hence, we investigate W $(A \vee B) \Rightarrow$ W $A \wedge$ W B by calculating equationally

$$\text{W}\,(A \vee B) \Rightarrow \text{W}\, A \;\equiv\; \langle\text{Exercise}\rangle \;\Box\Diamond B \Rightarrow \text{W}\, A \qquad (B.24)$$

and, therefore, W $(A \vee B) \Rightarrow$ W $A \wedge$ W $B \;\equiv\; (\Box\Diamond B \Rightarrow \text{W}\, A) \wedge (\Box\Diamond A \Rightarrow \text{W}\, B)$.

The r.h.s. is sufficient for W $(A \vee B) \Rightarrow$ W $A \wedge$ W B but also necessary and hence the weakest condition possible. Hence, the essential goal is amply met.

Just for completeness, one can investigate W $A \wedge$ W $B \Rightarrow$ W $(A \vee B)$ by calculating

W $A \Rightarrow$ W $B \Rightarrow$ W $(A \vee B)$

$\qquad \Leftarrow \langle\text{Exercise}\rangle \;\Box(\text{E}\, A \Rightarrow \Box\neg\text{E}\, B \vee \Diamond A) \wedge \Box(\text{E}\, B \Rightarrow \Box\neg\text{E}\, A \vee \Diamond B)$

$\qquad \equiv \langle\text{Introd. D}\rangle \;\text{D}\,(A, B) \wedge \text{D}\,(B, A).$

The operator D is defined by $\text{D}(A, B) \equiv \Box(\text{E}\, A \Rightarrow \Box\neg\text{E}\, B \vee \Diamond A)$.

Clearly, $DR1 \equiv \text{D}(\langle A\rangle_v, \langle B\rangle_v)$ and similarly $DR2 \equiv \text{D}(\langle B\rangle_v, \langle A\rangle_v)$.

Finally, one can check the relationship with (B.24) by calculating

$$\Box\Diamond B \Rightarrow \text{W}\, A \;\Leftarrow\; \langle\text{Exercise}\rangle \;\text{D}\,(A, B).$$

Many calculations were left as exercises to let the reader enjoy the feel of discovery.

B.5   Conclusions

Awareness in the use of model checking requires a higher mathematical standard than often suggested when advocating the use of tools. To make this standard more accessible, we have made the "user-friendliness" of the calculational style available in reasoning about specifications in general and temporal formulas in particular. To unify the various tool- or language-dependent temporal logics, a generic form (FTC) was used, which is pure predicate calculus. Specific logics are then captured in a very direct and simple way, illustrated in detail for TLA$^+$.

Note that Lamport's *Specifying Systems* (Lamport 2002a) concentrates on writing specifications. Proofs are considered in one chapter only, since introducing a temporal proof style (as in Manna and Pnueli (1991)) and meeting more proof obligations would have doubled the size of the book. Yet, not surprisingly, the proofs given concern patterns.

Reasoning about patterns helps keeping the complexity of temporal specifications manageable and within the grasp of intuition. The calculational approach makes this easier, and even supports discovery by newcomers in the field.

Still, in an educational environment, predicate calculus clearly remains a prerequisite, which is compensated by its very wide usefulness.

Appendix C
Comparision of Formal Methods

C.1 TLA$^+$ Model of Harris' Algorithm

```
 1 ┌──────────────────── MODULE HarrisR ────────────────────┐
   This model represents the algorithm presented by Harris for the implementation of non-
   blocking linked-lists.
   In this model the nodes inserted to the list are stored in the global variable "mem". The
   actions of deleting and inserting nodes to the list are divided, for a finer-grained solution in
   several intermediate steps.
   For 'artificial' (to be seen in some comments bellow) we mean all the aspects that are not
   directly connected to the algorithm, but that represent adaptations of it to be possible/more
   easily modeled.
10 EXTENDS Naturals, Sequences, FiniteSets
11 CONSTANT Adr,           the set of addresses
12           Keys,          the set of keys
13           Process,       the set of processes
14           HEAD, TAIL    values of the keys of 'Head' and 'Tail'

16 VARIABLES mem,        'state of the memory', with all the nodes inserted in the Linked List
17            proc,       auxiliary variable – information for the intermediate stages
18            setup       'artificial' variable for the initial insertion of 'Head' and 'Tail'

20 ASSUME   HEAD has to be smaller than any element of the set Keys, and TAIL has to
            be bigger
21              ∀ k ∈ Keys : HEAD < k  ∧  k < TAIL
22 ├─────────────────────────────────────────────────────────┤


25 TypeInvariant ≜  mem stores all the nodes inserted into the list. It's a function that assigns
26                  nodes to addresses. Each node as a key, a next field pointing to the
                    address of the next node, and can be marked or not.
28                  ∧ mem ∈ [Adr → [key : Keys ∪ {0, 1, 100},
29                                  next  : Adr ∪ {0},
30                                  mark : {0, 1}]]

32                  "setup" is 0 before the insertion of Head and Tail into the list, and 1
                    after that.
33                  ∧ setup ∈ {0, 1}
```

E. Verhulst et al., *Formal Development of a Network-Centric RTOS: Software
Engineering for Reliable Embedded Systems*, DOI 10.1007/978-1-4419-9736-4_11,
© Springer Science+Business Media, LLC 2011

35   "*proc*" is a function of each process
36                              $\land\ proc \in [$
37                                  $Process \rightarrow [$ *ninfo* is a record that keeps track of the
                                    information regarding
38   the current node to be inserted/deleted, like its key and position
                                    in the list.
39                                      $ninfo : [CNkey : Keys \cup \{0\}, CNnext :$
                                        $Adr \cup \{0\},$
40 $AdrLeft : Adr \cup \{0\}, AdrRight :$
40 $Adr \cup \{0\},$
41 $RigNext : Adr \cup \{0\}],$

43   *procIns* and *procDel* state the finer-grained
                                            steps of the insert
38   and delete actions
45                                      $procIns$: {"readyI", "createdI", "locatedI",
46   "unique , swapedI1"},

48                                      $procDel$: {"readyD", "identifiedD", "locatedD",
49   "assignedD", "swapedD1"},

51   A process can only start inserting/deleting a
                                            new node if it's
52   still free. Once it's committed to a certain action it should carry
                                    it till the
53   end. "choice" is used to represent that. A simple 2 bit variable,
                                    e.g. {"free",
54   "committed"}, would be a more elegant solution. The
                                    distinction between 'insert'
55   and 'delete' is necessary for 'artificial reasons' − see
                                    step "$CreateI(p, key)$".
56                                      $choice$: {"undecided", "toinsert", "todelete"}]]

59   $Coherence \stackrel{\Delta}{=}$ "The key of the node that one node points to has to smaller than its own key"
60   LET   set of all nodes pointing to another one
61                                  $nodp \stackrel{\Delta}{=} \{j \in Adr : (mem[j].key \neq 0 \land mem[j].next \neq 0\ )\}$
62   IN   Claim:
63                                  $\forall\, i \in nodp : mem[i].key < mem[mem[i].next].key$

66 ├──┤

Initially all memory is blank ('Head' and 'Tail' haven't been inserted) and the processes are
ready to start
71   $Init \stackrel{\Delta}{=}\ \land mem = [a \in Adr \mapsto [key \mapsto 0, next \mapsto 0, mark \mapsto 0]]$

73              $\land\ setup = 0$

75              $\land\ proc = [p \in Process \mapsto [$

77                          $ninfo \mapsto [CNkey \mapsto 0, CNnext \mapsto 0, AdrLeft \mapsto 0, AdrRight \mapsto 0,$
78                              $RigNext \mapsto 0],$

80                              $procIns \mapsto$ "readyl",

82                              $procDel \mapsto$ "readyD",

84                              $choice \mapsto$ "undecided"]]

87

88   Insertion of 'Head' and 'Tail' in the memory

90   $SetInitNodes \stackrel{\Delta}{=} \land setup = 0$

91   The Head is inserted in the first element of the memmory and the Tail in the second.

92                     $\land$ LET $fst \stackrel{\Delta}{=}$ CHOOSE $a \in Adr : \forall i \in Adr : a \leq i$

93                            $scd \stackrel{\Delta}{=}$ CHOOSE $a \in Adr \setminus \{fst\} : \forall i \in Adr \setminus \{fst\} : a \leq i$

94   IN

95                          $mem' = [mem$ EXCEPT $![fst].key = HEAD, ![fst].next = scd,$

96 $![scd].key = TAIL]$

97                     $\land setup' = 1$

98                     $\land$ UNCHANGED $\langle proc \rangle$

87

This is the beginning of the 'Insertion process'. It's divided in 5 steps: 'Createl', 'Locatel', 'VerUniq', 'CasI1' and 'CasI2'.

'Creation of the node' to insert to insert: a key (from the set of possible keys) is assigned as the node key.

109   $CreateI(p, key) \stackrel{\Delta}{=} \land setup = 1$

110                         $\land proc[p].procIns =$ "readyl"

111                         $\land proc[p].choice =$ "undecided"

112   Checks if there's still space in memmory

113                         $\land (Cardinality(\{a \in Adr : mem[a].key = 0\}) - Cardinality($

114                              $\{i \in Process : proc[i].choice =$ "toinsert"$\})) > 0$

116                         $\land proc' = [proc$ EXCEPT $![p].ninfo.CNkey = key,$

117 $![p].procIns =$ "createdl",

118 $![p].choice =$ "toinsert"]

119                      $\land$ UNCHANGED $\langle mem, setup \rangle$

'Location of the node' – The relative position of the node is determined and registered in "$AdrLeft$" and "$AdrRight$" - the addresses of the nodes that are, respectively, at the left and right of the new node. The order is determined by the value of "key" and the nodes are connected in ascending order.

125   $LocateI(p) \stackrel{\Delta}{=} \land proc[p].procIns =$ "createdl"

127                         $\land$ LET $elemr \stackrel{\Delta}{=} \{j \in Adr : (mem[j].key \neq 0 \land mem[j].mark = 0 \land$

128 $mem[j].key \geq proc[p]$
                                                   $.ninfo.CNkey)\}$

130                             $eleml \stackrel{\Delta}{=} \{j \in Adr : (mem[j].key \neq 0 \land mem[j].mark = 0 \land$

131 $mem[j].key < proc[p].ninfo$
                                                   $.CNkey)\}$

134
$$right \stackrel{\Delta}{=} \text{CHOOSE } a \in elemr : \forall i \in elemr : mem[a].key$$
$$\leq mem[i].key$$

136
$$left \stackrel{\Delta}{=} \text{CHOOSE } a \in eleml : \forall i \in eleml : mem[a].key$$
$$\geq mem[i].key$$

139
$$adj \stackrel{\Delta}{=} mem[left].next = right \quad \text{Check if nodes are adjacent}$$

141   IN
142   IF adj = TRUE THEN   proceed

144                        $\land proc' = [proc$ EXCEPT $![p].ninfo.AdrLeft = left,$
145 $![p].ninfo.AdrRight = right,$
146 $![p].procIns = $ "locatedl"]
147                        $\land$ UNCHANGED $\langle mem, setup \rangle$

149   ELSE   search again
150                        $\land proc' = [proc$ EXCEPT $![p].procIns = $ "createdl"]
151                        $\land$ UNCHANGED $\langle mem, setup \rangle$

'Verification of uniqueness' – Checks if the node already exists in the list. This is the case when the key value of the right node is the same as the key value of the node to be inserted

158 $VerUniq(p) \stackrel{\Delta}{=} \land proc[p].procIns = $ "locatedl"

160                        $\land$ IF $mem[proc[p].ninfo.AdrRight].key \neq proc[p].ninfo.CNkey$ THEN

162                        $proc' = [proc$ EXCEPT $![p].procIns = $ "unique"]   proceed

164   ELSE   If the key already exists, the process is aborted
165                        $proc' = [proc$ EXCEPT $![p].procIns = $ "readyl",
166 $![p].choice = $ "undecided"]

168                        $\land$ UNCHANGED $\langle mem, setup \rangle$

The current node is made to point to the right node:
173 $CasI1(p) \stackrel{\Delta}{=} \land proc[p].procIns = $ "unique"
174                        $\land proc' = [proc$ EXCEPT $![p].ninfo.CNnext = proc[p].ninfo.AdrRight,$
175 $![p].procIns = $ "swapedl1"]
176                        $\land$ UNCHANGED $\langle mem, setup \rangle$

The physical insertion of the node, recurring to CAS (addr, old, new), is attempted: the node that is being presently pointed at that instant by the left node [addr] is compared to the one previously identified [old], whose value has been stored in a variable and :
    – If they match, the left node is made to point to the current node, concluding the inserting process.
    – If they don't match (which means that some change in the list has been made in the mean time), the process goes back to LocateI).

185 $CasI2(p) \stackrel{\Delta}{=} \land proc[p].procIns = $ "swapedl1"

187      $\wedge$ IF $mem[proc[p].ninfo.AdrLeft].next = proc[p].ninfo.AdrRight$ THEN

190   LET $pos \stackrel{\Delta}{=}$ CHOOSE $a \in \{b \in Adr : mem[b].key = 0\}$:

191      $\forall i \in \{j \in Adr : mem[j].$

     $key = 0\} : a \leq i$

192   IN   insert in mem

193      $\wedge mem' = [mem$ EXCEPT $![pos].key = proc[p].ninfo.CNkey,$

194      $![pos].next = proc[p]$

     $.ninfo.CNnext,$

195      $![proc[p].ninfo.AdrLeft]$

     $.next = pos]$

197   and Reset the procedure

198      $\wedge proc' = [proc$ EXCEPT $![p].ninfo.CNkey = 0,$

199      $![p].ninfo.CNnext = 0,$

200      $![p].ninfo.AdrLeft = 0,$

201      $![p].ninfo.AdrRight = 0,$

202      $![p].ninfo.RigNext = 0,$

203      $![p].procIns =$ "readyI",

204      $![p].choice =$ "undecided"]

206      $\wedge$ UNCHANGED $\langle setup \rangle$

209   ELSE   search again

211      $\wedge proc' = [proc$ EXCEPT $![p].procIns =$ "createdI"]

213      $\wedge$ UNCHANGED $\langle mem, setup \rangle$

End of the insertion process

219 ├───┤

This is the beginning of the 'Deletion process'. It's divided in 5 steps: 'Identify', 'LocateD', 'AssignD', 'CasD1' and 'CasD2'.

'Creaction of the node to insert' – an identifier key is selected

226 $Identify(p, key) \stackrel{\Delta}{=} \wedge setup = 1$

227      $\wedge proc[p].procDel =$ "readyD"

228      $\wedge proc[p].choice =$ "undecided"

229      $\wedge proc' = [proc$ EXCEPT $![p].ninfo.CNkey = key,$

230      $![p].procDel =$ "identifiedD",

231      $![p].choice =$ "todelete"]

232      $\wedge$ UNCHANGED $\langle mem, setup \rangle$

The list is searched in order to identify the left and right nodes: the left is the unmarked node that has the biggest key strictly smaller; the right is the unmarked node that has the smallest key greater or equal (than the current's node key).

237 $LocateD(p) \stackrel{\Delta}{=} \wedge proc[p].procDel =$ "identifiedD"

239      $\wedge$ LET $posr \stackrel{\Delta}{=} \{j \in Adr : (mem[j].key \neq 0 \wedge mem[j].mark = 0 \wedge$

240      $mem[j].key \geq proc[p]$

     $.ninfo.CNkey)\}$

242        $posl \stackrel{\Delta}{=} \{j \in Adr : (mem[j].key \neq 0 \wedge mem[j].mark = 0 \wedge$
243 $mem[j].key < proc[p]$
                                              $.ninfo.CNkey)\}$

245        $right \stackrel{\Delta}{=} \text{CHOOSE } a \in posr : \forall i \in posr : mem[a]$
           $.key \leq mem[i].key$

247        $left \stackrel{\Delta}{=} \text{CHOOSE } a \in posl : \forall i \in posl : mem[a]$
           $.key \geq mem[i].key$

249   IN
250   Check the identity of the right node: if its key is equal to the current one the process
251   continuous to execute and the right node is the node to be deleted; otherwise it aborts.
252   IF $mem[right].key = proc[p].ninfo.CNkey$ THEN

254                    $\wedge proc' = [proc \text{ EXCEPT } ![p].ninfo.AdrLeft = left,$
255 $![p].ninfo.AdrRight = right,$
256 $![p].procDel = \text{"locatedD"}]$
257                    $\wedge \text{UNCHANGED } \langle mem, setup \rangle$

259   ELSE   abort (key doesn't exist)
260                    $\wedge proc' = [proc \text{ EXCEPT } ![p].ninfo.CNkey = 0,$
261 $![p].procDel = \text{"readyD"},$
262 $![p].choice = \text{"undecided"}]$
263                    $\wedge \text{UNCHANGED } \langle mem, setup \rangle$

The immediate successor of the right node is stored in 'RigNext'
268 $AssignD(p) \stackrel{\Delta}{=} \wedge proc[p].procDel = \text{"locatedD"}$
269                          $\wedge proc' = [proc \text{ EXCEPT } ![p].ninfo.RigNext = mem[proc[p]$
                              $.ninfo.AdrRight].next,$
270 $![p].procDel = \text{"assignedD"}]$
271                    $\wedge \text{UNCHANGED } \langle mem, setup \rangle$

Marking of the node: if the right node next field is still pointing to the node previously identified, then the node is marked. If not the process goes back to 'LocateD'.
276 $CasD1(p) \stackrel{\Delta}{=} \wedge proc[p].procDel = \text{"assignedD"}$
277   The operation is guarded by the pre-condition that $RigNext$ is not a marked node; otherwise the process goes directly back to 'LocateD'
278
279                    $\wedge \text{IF } mem[proc[p].ninfo.RigNext].mark = 0$ THEN

281   IF $mem[proc[p].ninfo.AdrRight].next = proc[p].ninfo$
                              $.RigNext$ THEN

283                          $\wedge \quad mem' = [mem \text{ EXCEPT } ![proc[p].ninfo.AdrRight]$
                              $.mark = 1]$
284                          $\wedge \quad proc' = [proc \text{ EXCEPT } ![p].procDel = \text{"swapedD1"}]$
285                          $\wedge \quad \text{UNCHANGED } \langle setup \rangle$

287   ELSE   search again
288                                $\wedge proc' = [proc \text{ EXCEPT } ![p].procDel = \text{"identifiedD"}]$

289                                 $\wedge$ UNCHANGED $\langle mem, setup \rangle$

291   ELSE   search again
292                         $\wedge\, proc' = [proc$ EXCEPT $![p].procDel =$ "identifiedD"]
293                         $\wedge$ UNCHANGED $\langle mem, setup \rangle$

Removal of the node from the linked-list: if the left node next field is still pointing to the right node, then it is made to point to the node identified as its immediate successor. If the comparison fails, the process goes back to 'LocateD'.

298 $CasD2(p) \overset{\Delta}{=} \wedge\, proc[p].procDel =$ "swapedD1"

300                         $\wedge$ IF $mem[proc[p].ninfo.AdrLeft].next = proc[p].ninfo.AdrRight$ THEN

302                             $\wedge\, mem' = [mem$ EXCEPT $![proc[p].ninfo.AdrLeft].next = proc[p]$
                                $.ninfo.RigNext]$

304                             $\wedge\, proc' = [proc$ EXCEPT $![p].ninfo.CNkey = 0,$
305 $![p].ninfo.CNnext = 0,$
306 $![p].ninfo.AdrLeft = 0,$
307 $![p].ninfo.AdrRight = 0,$
308 $![p].ninfo.RigNext = 0,$
309 $![p].procDel =$ "readyD",
310 $![p].choice =$ "undecided"]

312                             $\wedge$ UNCHANGED $\langle setup \rangle$

314   ELSE   search again
315                             $\wedge\, proc' = [proc$ EXCEPT $![p].ninfo.AdrLeft = 0,$
316 $![p].ninfo.AdrRight = 0,$
317 $![p].ninfo.RigNext = 0,$
318 $![p].procDel =$ "identifiedD"]
319                             $\wedge$ UNCHANGED $\langle mem, setup \rangle$

End of the deletion process
324

"Summary" of the specification

329 $Insert(i, k) \overset{\Delta}{=} CreateI(i, k) \vee LocateI(i) \vee VerUniq(i) \vee CasI1(i) \vee CasI2(i)$

331 $Delete(i, k) \overset{\Delta}{=} Identify(i, k) \vee LocateD(i) \vee AssignD(i) \vee CasD1(i) \vee CasD2(i)$

334 $Next \overset{\Delta}{=} \vee\, SetInitNodes$

336                 $\vee\, \exists i \in Process : \exists k \in Keys : Insert(i, k) \vee Delete(i, k)$

339 $Spec \overset{\Delta}{=} Init \wedge \Box[Next]_{\langle mem, proc, setup \rangle}$
340
341 THEOREM $Spec \implies \Box(TypeInvariant \wedge Coherence)$
342

C.2   Promela Model of Harris' Algorithm

```
 1   #define L        4      /* Lenght of the memmory */
 2   #define M        2      /* Number of inserting processes */
 3   #define N        2      /* Number of deleting processes */
 4   #define HEAD     1      /* Key value of the 'Head' */
 5   #define TAIL   100      /* Key value of the 'Tail' */
 6   typedef Node {
 7     byte key;
 8     byte next;
 9     bool mark
10   }
11   Node mem[L];
12   bool setdone=false
13   /*|========|=========|=========|=========|=========| */
14   /*       setup - Insertion of 'Head' and 'Tail'         */
15   active proctype setup()
16   {
17     if
18     ::setdone == false ->
19        atomic{
20          mem[0].key=HEAD;
21          mem[0].next=1;
22          mem[1].key=TAIL;
23          setdone=true;
24        }
25     ::else->skip
26     fi
27   }
28   /* ========|=========|=========|=========|=========|
29           Insert process
30   ========|=========|=========|=========|=========| */
31   active [M] proctype insert()
32   {
33   (setdone==true);   /* Guarding condition for the executability of
34                          any inserting process */
35                       /* - the initial setup has to be finished firt
36                         */\ \
37   byte CNkey, CNnext, AdrLeft, AdrRight, t, t_next,  left_next;
38   byte pos=0, counter=0, nonblank=0;\
39   startinsert:\
40   /* "CreateI - Ii" */\
41   atomic{\
42      do    /* count the elements in the list */
43      :: (mem[counter].key != 0 \&\& mem[counter].key < TAIL) ->
44         nonblank ++;
45         counter = mem[counter].next;
46      :: else -> break
47      od;\
48      if /* an insertion can only happen if there's space in mem */
49      ::((L - nonblank - M) > 0 ) ->
50
```

```
51          if
52          :: CNkey=10
53          :: CNkey=20
54          :: CNkey=30
55          :: CNkey=40
56          :: CNkey=50
57          fi
58      :: else -> goto endinsert
59      fi
60  }\ \
61  /*   "LocateI - IIi"   */\
62  searchagain:\
63      t=0;
64      t_next=mem[0].next;
65      left_next=0;
66      pos=0;\ \
67      do
68      :: ((mem[t_next].mark==true) || (mem[t].key < CNkey)) ->\
69          if
70          :: (mem[t_next].mark==false) ->
71              AdrLeft = t;
72              left_next = t_next;
73          :: else ->skip
74          fi;\
75          t=t_next;\
76          if
77          :: (t==1) -> goto endcycle
78          :: else ->skip
79          fi;\
80          t_next = mem[t].next;\
81      :: else -> break
82      od;\ \
83  endcycle:
84      AdrRight=t;\ \
85      if
86      :: (left_next == AdrRight) ->
87          if
88          :: ((AdrRight!=1) $\&\&$ (mem[mem[AdrRight].next].
89                  mark == true)) ->
90                  goto searchagain
91          :: else -> goto endsearch
92          fi
93      :: else -> goto endsearch
94      fi;\ \
95  endsearch:
96      skip;\ \
97  /*   "VerUniq - IIIi"   */\
98      if
99      :: ((AdrRight!=1) $\&\&$ (mem[AdrRight].key == CNkey))
100         -> goto endinsert
101     :: else -> skip
102     fi;\ \
103 /*   "CasI1 - IVi"   */\
```

```
104        CNnext = AdrRight;\ \
105  /* "CasI2 - Vi"  */\
106  atomic{
107        do
108        :: (mem[pos].key != 0) -> pos++
109        :: else -> break
110        od;
111  }\ \
112        if
113        :: (mem[AdrLeft].next == AdrRight) ->
114            mem[AdrLeft].next = pos;
115            mem[pos].key = CNkey;
116            mem[pos].next = CNnext
117        :: else -> goto searchagain
118        fi;\
119  endinsert:
120        skip;\ \
121  } /*  End of proctyte insert    */ \ \
122  /* ========|=========|=========|=========|=========|
123
124                            Delete process
125
126  ========|=========|=========|=========|=========| */\ \
127  active [N] proctype delete()
128  {
129  (setdone==true); /* Guarding condition for the executability of
130    any deleting process - the initial setup has to be finished
131    first */\ \
132  byte CNkey, AdrLeft, AdrRight, RigNext, pos;
133  byte t=0;
134  byte t_next=mem[0].next;
135  byte left_next=0;\
136  /* Identify - Id */\
137        if
138        :: CNkey=10
139        :: CNkey=20
140        :: CNkey=30
141        :: CNkey=40
142        :: CNkey=50
143        fi;\ \
144  /* LocateD - IId */\
145  searchagainD:\
146        t=0;
147        t_next=mem[0].next;
148        left_next=0;
149        pos=0;\ \
150        do
151        :: ((mem[t_next].mark==true) || (mem[t].key < CNkey)) ->
152            if
153            :: (mem[t_next].mark==false) ->
154                AdrLeft = t;
155                left_next = t_next;
156            :: else ->skip
```

```
157          fi;\
158          t=t_next;\
159          if
160          :: (t==1) -> goto endcycleD
161          :: else ->skip
162          fi;\
163          t_next = mem[t].next\
164      :: else -> break
165      od;\ \
166  endcycleD:
167      AdrRight=t;\ \
168      if
169      :: (left_next == AdrRight) ->
170          if
171          :: ((AdrRight!=1) $\&\&$ (mem[mem[AdrRight].next]
172             .mark == true)) ->
173                     goto searchagainD
174          :: else -> goto endsearchD
175          fi
176      :: else -> goto endsearchD
177      fi;\ \
178  endsearchD:
179      skip;\ \
180  /* Examine - IIId */\
181      if
182      :: ((AdrRight!=1) || (mem[AdrRight].key != CNkey)) -> goto
183        enddelete
184      :: else -> skip
185      fi;\ \
186  /* AssignD - IVd */\
187      RigNext = mem[AdrRight].next;\ \
188  /* CasD1 - Vd */\
189      if
190      :: (mem[RigNext].mark == false) ->
191          if
192          :: (mem[AdrRight].next == RigNext) -> mem[AdrRight].
193        mark = true
194          :: else -> goto searchagainD
195          fi;
196      :: else -> goto searchagainD
197      fi;\ \
198  /* CasD2 - VId */\
199      if
200      :: (mem[AdrLeft].next == AdrRight) -> mem[AdrLeft].next
201        == RigNext
202      :: else -> goto searchagainD
203      fi;
204  enddelete:
205      skip;
206  } /* End of proctyte delete */ \ \
207  /* ========|=========|=========|=========|=========|
208  Correctness Claim
209  ========|=========|=========|=========|=========| */
```

```
210    never {
211      do
212      :: !( !( (mem.key != 0) \&\& (mem.next != 0) ) || (mem.key <
213       mem[mem.next].key)) -> break
214      :: else
215      od
216    }
217    /* End of model */
```

Listing C.1 Configuration file

Glossary

CPU	Central Processing Unit. 160–163
CSP	Communicating Sequential Processes. 163
Event	An Event Hub synchronises Tasks using a boolean. 107, 109
FIFO	A FIFO Hub buffers and transfers data between Tasks in FIFO order.. 107, 109
Hub	A Hub is the interaction entity used by Tasks to cooperate. 107, 108
IRQ	Interrupt Request. 143, 160–163, 165, 166, 168
ISR	Interrupt Service Routine. 143, 161–163, 165, 166, 168
LTL	linear temporal logic. 61, 70, 71
Memory Pool	A Memory Pool Hub manages a Pool of Memory Blocks. 107, 109
NMI	Non Maskable Interrupt. 161
OS	Operating System. 161, 162
Packet	A Packet is the standardised datastructure used in OpenComRTOS to implement the services. 107
Packet Pool	A Packet Pool Hub manages a Pool of Packets. 109
Port	A Port Hub allows to transfer Packets between Tasks. 109
Resource	A Resource Hub provides exclusive access to a logical resource. 107, 109
RTOS	Real Time Operating System. 143, 161
Semaphore	A Semaphore Hub synchronises Tasks using a counter. 107, 109
Task	A function with its own workspace (stack) and priority scheduled in OpenComRTOS. 107
UART	Universal Asynchronous Receiver/Transmitter. 160

E. Verhulst et al., *Formal Development of a Network-Centric RTOS: Software Engineering for Reliable Embedded Systems*, DOI 10.1007/978-1-4419-9736-4, © Springer Science+Business Media, LLC 2011

References

Ban (2003). Bandera home page (2003). http://bandera.projects.cis.ksu.edu/. [Online; accessed 1-December-2010].

ALT (2010). Altreonic. http://www.altreonic.com, last visited: 20.01.2011.

IEE (2011). Ieee std 1355-1995 standard for heterogeneous interconnect (hic). http://grouper.ieee.org/groups/1355/index.html, last visited: 20.01.2011.

INM (2011). Inmos. http://www.inmos.com, last visited: 20.01.2011.

MAS (2011). Mast. http://mast.unican.es, last visited: 20.01.2011.

OLS (2011). Ols. http://www.openlicensesociety.org, last visited: 20.01.2011.

Spa (2011). Spacewire. http://spacewire.esa.int, last visited: 20.01.2011.

XMO (2011). Xmos. http://www.xmos.com, last visited: 20.01.2011.

Booch, G., Maksimchuk, R. A., Engel, M. W., Young, B. J., Conallen, J., and Houston, K. A. (2007). *Object-Oriented Analysis and Design with Applications*. Addison-Wesley, third edition.

Boute, R. (1986). A calculus for reasoning about temporal phenomena. In *Proceedings 4th NGI-SION Symposium*, pages 405–411.

Boute, R. (2005). Functional declarative language design and predicate calculus: a practical approach. *ACM Trans. Program. Lang. Syst.*, 27(5):988–1047.

Boute, R. (2006a). Microsemantics as a bootstrap in teaching formal methods. In Boca, P. and Duce, D., editors, *Teaching Formal Methods: Practice and Experience*. http://www.bcs.org/server.php?show=conWebDoc.9094.

Boute, R. (2006b). Using domain-independent problems for introducing formal methods. In Misra, J., Nipkow, T., and Sekerinski, E., editors, *FM 2006: Formal Methods*, volume 4085 of *Lecture Notes in Computer Science / Programming and Software Engineering*. Springer.

Boute, R. and Verlinde, H. Functionals for the semantic specification of temporal formulas for model checking. In Hartmut König, Monika Heiner, A. W., editor, *FORTE 2003 Work-in-Progress Papers*.

Boute, R. T. (2003). Concrete generic functionals: Principles, design and applications. In Gibbons, J. and Jeuring, J., editors, *Generic Programming*, volume 115 of *IFIP Advances in Information and Communication Technology*, pages 89–119. Springer.

Boute, R. T. (2006c). Calculational semantics: Deriving programming theories from equations by functional predicate calculus. *ACM Trans. Program. Lang. Syst.*, 28(4):747–793.

Briand, L. P. and Roy, D. M. (1999). *Meeting Deadlines in Hard Real-Time Systems: The Rate Monotonic Approach*. IEEE.

Clarke, E. M., Grumberg, O., and Peled, D. A. (1999). *Model Checking*. The MIT Press.

Cousot, P. (2008). Abstract Interpretation. http://www.di.ens.fr/~cousot/AI/.

Davies, J. and Schneider, S. (1989). Introduction to Timed CSP, University of Oxford Computing Lab, Programming Research Gp. ISBN: 0902928570.

E. Verhulst et al., *Formal Development of a Network-Centric RTOS: Software Engineering for Reliable Embedded Systems*, DOI 10.1007/978-1-4419-9736-4,
© Springer Science+Business Media, LLC 2011

Dijkstra, E. W. (1975). Guarded commands, nondeterminacy and formal derivation of programs. *Commun. ACM*, 18(8):453–457.

Dijkstra, E. W. (1989). To hell with "meaningful identifiers"! circulated privately.

Dijkstra, E. W. (1990). How Computing Science created a new mathematical style. *EWD 1073* http://www.cs.utexas.edu/users/EWD/ewd10xx/EWD1073.PDF.

Dijkstra, E. W. (1997). *A Discipline of Programming*. Prentice Hall PTR, Upper Saddle River, NJ, USA.

Dijkstra, E. W. and Scholten, C. S. (1990). *Predicate calculus and program semantics*. Springer-Verlag New York, Inc., New York, NY, USA.

Dong, J. S., Zhang, X., Sun, J., and Hao, P. (2006). Reasoning about timed csp models. Technical report, School of Computing, National University of Singapore.

Dwyer, M. B., Avrunin, G. S., and Corbett, J. C. (1998). Property specification patterns for finite-state specification. In *Proceedings of FMSP'98, Second Workshop on Formal Methods in Software Practice*.

Dwyer, M. B., Avrunin, G. S., and Corbett, J. C. (1999). Patterns in property specications for finite-state verication. In *In Proceedings of the 21st International Conference on Software Engineering ICSE'99*.

Dwyer, M. B. and Hatcliff, J. (2002). Bandera temporal specification patterns. internet. tutorial presentation at ETAPS'02 (Grenoble) and SMF'02 (Bertinoro), http://santos.cis.ksu.edu/bandera/Talks/SFM02/02-SFM-Patterns.ppt.

Gao, H. and Hesselink, W. H. (October 2004). A general lock-free algorithm using compare-and-swap.

Gries, D. (1991). Improving the curriculum through the teaching of calculation and discrimination. *Communications of the ACM*, 34(3):45–55.

Gries, D. and Schneider, F. B. (1993). *A Logical Approach to Discrete Math*. Springer, 1st edition.

Habrias, H. and Faucou, S. (2004). Linking paradigms, semi-formal and formal notations. In Dean, C. N. and Boute, R. T., editors, *Teaching Formal Methods*, volume 3294 of *Lecture Notes in Computer Science*, pages 166–184. Springer. CoLogNET/FME Symposium, TFM 2004, Ghent, Belgium, November 18-19, 2004.

Harbour, M. G., Medina, J., Gutirrez, J., Palencia, J., and Drake., J. (2002). Mast: An open environment for modeling, analysis, and design of real-time systems. 1st carts workshop, aranjuez, spain, october 2002.

Harris, T. L. (2001). A pragmatic implementation of non-blocking linked-lists. In *DISC '01: Proceedings of the 15th International Conference on Distributed Computing*, pages 300–314, London, UK. Springer-Verlag.

Herlihy, M. (1993). A methodology for implementing highly concurrent objects. *ACM Trans. Program. Lang. Syst.*, 15(5):745–770.

Hoare, C. (1985a). *C.A.R. Hoare. Communicating Sequential Processes*. Prentice-Hall.

Hoare, C. A. R. (1985b). *Communicating sequential processes*. Prentice-Hall, Inc., Upper Saddle River, NJ, USA.

Holzmann, G. (2003a). *The SPIN Model Checker: Primer and Reference Manual*. Pearson Education.

Holzmann, G. (2003b). *Spin model checker, the: primer and reference manual*. Addison-Wesley Professional.

IBM (1983). IBM System/370 Extended Architecture, Principles of Operation. Publication No. SA22-7085.

Intel Corporation (2002, Revision 2.1). *Intel Itanium Architecture Software Developer's Manual. Volume 1: Application Architecture*.

Jayanti, P. and Petrovic, S. (2004). Efficient Wait-Free Implementation of Multiword LL/SC Variables. Technical Report TR2004-523, Dartmouth College, Computer Science, Hanover, NH.

Klein, M., Ralya, T., Pollak, B., Obenza, R., and Harbour, M. G. (1993). *A Practitioner's Handbook for Real-Time Analysis: Guide to Rate Monotonic Analysis for Real-Time Systems*. Springer.

Lamport, L. (1983). Specifying concurrent program modules. *ACM Trans. Program. Lang. Syst.*, 5(2):190–222.

Lamport, L. (2002a). *Specifying Systems: The TLA+ Language and Tools for Hardware and Software Engineers*. Addison-Wesley Longman Publishing Co., Inc., Boston, MA, USA.

Lamport, L. (2002b). *Specifying Systems: The TLA+ Language and Tools for Hardware and Software Engineers*. Addison-Wesley Longman Publishing Co., Inc., Boston, MA, USA.

Liu, C. L. and Layland, J. W. (1973). Scheduling algorithms for multiprogramming in a hard-real-time environment. *J. ACM*, 20:46–61.

Manna, Z. and Pnueli, A. (1991). *The Temporal Logic of Reactive and Concurrent Systems – Specification*. Springer, 1st edition.

Michael, M. (2004a). Practical lock-free and wait-free LL/SC/VL implementations using 64-bit cas. In Guerraoui, R., editor, *Distributed algorithms*, volume 3274/2004 of *Lecture Nodes in Computer Science*, pages 144–158.

Michael, M. M. (2002a). High performance dynamic lock-free hash tables and list-based sets. In *SPAA '02: Proceedings of the fourteenth annual ACM symposium on Parallel algorithms and architectures*, pages 73–82, New York, NY, USA. ACM Press.

Michael, M. M. (2002b). Safe memory reclamation for dynamic lock-free objects using atomic reads and writes. In *PODC '02: Proceedings of the twenty-first annual symposium on Principles of distributed computing*, pages 21–30, New York, NY, USA. ACM Press.

Michael, M. M. (2004b). ABA prevention using single-word instructions. Technical Report RC23089, IBM Research Division.

Moir, M. (1997). Practical implementations of non-blocking synchronization primitives. In *PODC '97: Proceedings of the sixteenth annual ACM symposium on Principles of distributed computing*, pages 219–228, New York, NY, USA. ACM Press.

Roscoe, A. W. (1998). *The Theory and Practice of Concurrency*. Pearson Education Limited, Essex, CM20 2JE, England.

Sifakis, J. (2010). Formal methods and their evaluation. www-verimag.imag.fr/~sifakis/RECH/ FEMSYS/paper.ps. [Online; accessed 1-December-2010].

SPARC International (1994). *The SPARC architecture manual: Version 9*. Prentice-Hall.

Styenko, A. (1985). *Real-Time Systems: Scheduling and Structure Af.Sc. Thesis*. University of Toronto.

Verhulst, E. (1993a). Beyond the von neumann machine: communication as the driving design paradigm for mp-soc from software to hardware.

Verhulst, E. (1993b). Virtuoso : providing sub-microsecond context switching on dsps with a dedicated nanokernel. in international conference on signal processing applications and technology, santa clara september, 1993.

Verhulst, E. (1997a). Beyond transputing : fully distributed semantics in Virtuoso's Virtual Single Processor programming model and it's implementation on of-the-shelf parallel DSPs. In Bakkers, A. W. P., editor, *Proceedings of WoTUG-20: Parallel Programming and Java*, pages 77–86.

Verhulst, E. (1997b). Non-sequential processing: bridging the semantic gap left by the von neumann architecture. In *Signal Processing Systems SIPS'97*, pages 35–49.

Verhulst, E. (2002). The rationale for distributed semantics as a topology independent embedded systems design methodology and its implementation in the virtuoso rtos. *Design Automation for Embedded Systems*, 6:277–294. 10.1023/A:1014018820691.

Verhulst, E. and De Jong, G. (2007). Opencomrtos: an ultra-small network centric embedded rtos designed using formal modeling. In *Proceedings of the 13th international SDL Forum conference on Design for dependable systems*, SDL'07, pages 258–271, Berlin, Heidelberg. Springer-Verlag.

Verhulst, E., de Jong, G., and Mezhuyev, V. (2008). An industrial case: Pitfalls and benefits of applying formal methods to the development of a network-centric rtos. In Cuellar, J., Maibaum, T., and Sere, K., editors, *FM 2008: Formal Methods*, volume 5014 of *Lecture Notes in Computer Science*, pages 411–418. Springer Berlin / Heidelberg. 10.1007/978-3-540-68237-0.29.

Wikipedia (2011). Transputer – wikipedia, the free encyclopedia. [Online; accessed 1-February-2011].

Index

E. Verhulst et al., *Formal Development of a Network-Centric RTOS: Software Engineering for Reliable Embedded Systems*, DOI 10.1007/978-1-4419-9736-4,
© Springer Science+Business Media, LLC 2011